Lecture Notes in Applied and Computational Mechanics

Volume 33

Series Editors

Prof. Dr.-Ing. Friedrich Pfeiffer
Prof. Dr.-Ing. Peter Wriggers

Lecture Notes in Applied and Computational Mechanics

Edited by F. Pfeiffer and P. Wriggers

Further volumes of this series found on our homepage: springer.com

Spectral Method in Multiaxial Random Fatigue

Adam Niesłony · Ewald Macha

With 57 Figures and 7 Tables

 Springer

A. Niesłony
E. Macha
Opole University of Technology
Faculty of Mechanics
Department of Mechanics and Machine Design
ul. Mikolajczyka 5
45-271 Opole, POLAND
a.nieslony@po.opole.pl
e.macha@po.opole.pl

ISBN 978-3-642-09304-3   e-ISBN 978-3-540-73823-7

ISSN 1613-7736

Springer is a part of Springer Science+Business Media
springer.com
© Springer-Verlag Berlin Heidelberg 2007
Softcover reprint of the hardcover 1st edition 2007

Cover design: Künkel & Lopka, Heidelberg

Preface

The monograph is devoted to the spectral method for determination of fatigue life of machine elements and structures under multiaxial service loading. A detailed review of literature on spectral methods that includes: publications describing various relations between histories of loading and power spectral density functions, some methods of fatigue life determination under Gaussian loading and under multiaxial service loading has been presented in the monograph. The monograph contains theoretical foundations of the spectral method for fatigue life determination. The authors have discussed a rule of description of random loading states with the matrix of power spectral density functions of the stress/strain tensor components. Some chosen criteria of multiaxial fatigue failure being linear combinations of stress or strain components on the critical plane have been analyzed. The proposed formula enables to determine power spectral density of the equivalent history directly from the components of the power spectral density matrix of the multidimensional stochastic process. The assumptions and the procedure of determination of basic relationships of the spectral method according to stress and strain approaches have been presented. The authors worked out equations determining the fatigue life according to the spectral method using various linear hypotheses of fatigue damage accumulation. The proposed algorithm of fatigue life contains five blocks: 1 – determination of loading, 2 – determination of the critical plane position for the assumed multiaxial fatigue failure criterion, 3 – determination of power spectral density of the equivalent stress or strain, 4 – determination of statistical parameters of the equivalent parameter responsible for fatigue damage, and 5 – fatigue life calculation according to a suitable hypothesis of damage accumulation. In order to compare the results of fatigue life estimations many simulation tests were performed with the use of the spectral and the cycle counting methods. The comparison was done in three stages: 1 – positions of critical planes determined with the variance method in frequency and time domains, 2 – some chosen statistical parameters of the equivalent histories (distributions of instantaneous values, distributions of amplitudes after cycle counting and power spectral density functions), 3 –

lives calculated with the spectral and the cycle counting methods. The monograph contains the results of the test for 18G2A steel under random bending, torsion and combined bending with torsion with different correlations and levels of torsional and bending loading components. The results from the test and from the calculations with the use of spectral method and the cycle counting method have been compared and discussed. It appears that both considered methods give similar and satisfactory estimations of fatigue life. We hope that problems discussed in the monograph are interesting for experienced structure integrity engineers and give them possibility to broaden the knowledge about the efficient methods of lifetime evaluation of machine elements and structures under multiaxial random loading. The book can be also recommended to postgraduate and PhD students with an interest in fatigue of engineering materials.

A large part of the editorial works connected with the book was made during the Alexander von Humboldt Research Fellowships in Germany. Dr. Adam Niesłony wants to thank the AvH Foundation for their support as well as Prof. Sonsino from Fraunhofer Institute for Structural Durability and System Reliability LBF in Darmstadt for scientific and personal care during the stay in Germany. Authors also want to thank the Commission of the European Communities for the financial support as part of the CESTI project under the FP5 GROWTH Programme, contract No. G1MA-CT-2002-04058. Special thanks are dedicated to friends from the Department of Mechanics and Machine Design, Opole University of Technology, for the valuable help in working the book out, particularly to Prof. Tadeusz Łagoda, Dr. Aleksander Karolczuk and Dr. Henryk Achtelik.

Adam Niesłony
Ewald Macha

Department of Mechanics and Machine Design
Mechanical Engineering Faculty
Opole University of Technology
ul. Mikolajczyka 5
45-271 Opole, Poland

Contents

List of Symbols

A	—	coefficient derived from Wöhler fatigue characteristic,
b	—	fatigue strength exponent,
c	—	fatigue strain exponent,
D	—	damage,
E	MPa	Young modulus,
$\mathrm{e}\,[\cdot]$	—	expected value of an expression,
$\hat{\cdot}$	—	mean value, expected value of a variable,
f	s^{-1}	frequency,
f_0	s^{-1}	dominant frequency,
F	N	force,
$G(f)$	—	one-sided power spectral density function,
$G^*(f)$	—	complex conjugate of the function $G(f)$,
$h(\tau)$	—	impulse response (transfer) function of physical system,
j	—	imaginary unit,
I	—	irregularity coefficient,
m	—	exponent of the Wöhler fatigue characteristic,
m_k	—	k-th moment of spectral power density function,
M	Nm	moment of force,
M^+	—	expected number of peaks (local extremes) in a time unit,
N_0^+	—	expected number transitions of zero level with positive slope in a time unit,
N_f	cycles	number of cycles to failure,
N_G	cycles	number of cycles to failure for fatigue limit,
p	—	probability density,
P	—	probability,
r	—	correlation coefficient,
$R(\tau)$	—	correlation function,
$\mathrm{sgn}(\cdot)$	—	signum function,
$t,\ \tau$	s	time,

T	s	period,
σ	MPa	stress,
$\boldsymbol{\sigma}(t)$	MPa	stress tensor,
$\boldsymbol{\varepsilon}(t)$	—	strain tensor,
σ_{af}	MPa	fatigue strain limit for tension-compression,
τ_{af}	MPa	fatigue strain limit for torsion,
σ'_f	MPa	fatigue strength coefficient for tension-compression,
ε'_f	—	fatigue plastic strain coefficient,
ε	—	strain,
ϵ	—	spectrum width parameter,
κ	—	kurtosis,
μ	—	covariance,
$\Gamma(\cdot)$	—	gamma function,
$\Gamma(\cdot,\cdot)$	—	incomplete gamma function,
ω	$\mathrm{rad} \cdot \mathrm{s}^{-1}$	angular frequency (pulsatance),
Δ	—	range,
λ	—	coefficient including spectrum width,
γ	—	engineering shear strain.

Indices and Abbreviations

a	–	amplitude,
m	–	mean value,
ZC	–	cycle counting method,
SP	–	spectral method,
min	–	minimum,
max	–	maximum,
eq	–	equivalent or reduced value,
PSD	–	power spectral density,
FFT	–	fast Fourier transformation,
IFFT	–	inverse fast Fourier transformation,
NB	–	narrow-band frequency spectrum,
BB	–	broad-band frequency spectrum,
RC	–	range counting,
RF	–	rain flow,
RFC	–	rain flow counting.

1

Introduction

The question of material, machine and construction fatigue occurs prevalently in machine, building, maritime, and aircraft industries both at the stage of manufacturing and during operation. Some structural components suffer from failure associated with fatigue in the conditions of cyclic or random multiaxial loading. The fatigue process is particularly hazardous in aviation, power engineering and multiple supporting structures (e.g. jack-up platforms, bridges, towers, masts), in which the failure of a component causes huge material losses, ecological disaster and is directly threatening to the life and health of people. It is indicated [3, 11, 14, 63] that a number of random loading following natural phenomena (sea waves, variable wind velocity, etc.) is Gaussian in character with various spectral frequency widths. Therefore, this case is particularly taken into account in this book.

Research throughout the world is devoted to an elaboration of a universal method for determination of machine and component strength. A calculation algorithm with an application to a wide spectrum of materials under random loading, starting with a simple uniaxial state and involving complex multiaxial loading states is attempted in the course of research worldwide. It could be indicated that calculation algorithms in bibliography are divided into a group of cycle counting method, in which loading is analysed in the time domain, and a group of spectral methods in which power spectral density function defined in the frequency domain is applied for the description of loading.

In the cycle counting methods, the assessment of fatigue related strength is enabled by adequate criteria of multiaxial random fatigue, which are applied for determination of the equivalent stress or strain histories [2, 36]. Subsequently, algorithms for extracting cycles of histories and familiar procedures for accumulation of damage related failure stated for uniaxial random fatigue are applied [22, 95]. The postulated algorithm of determination of fatigue life by means of the spectral method applies the fatigue failure criteria and consists in the reduction of the multiaxial loading state to an equivalent uniaxial state in the frequency domain. By reduction of the multiaxial loading state the multiaxial fatigue assessment criteria based on critical plane concept was used. For determination of the necessary critical plane position the methods of variance or damage accumulation can be applied. Statistical parameters applicable for a description of the amplitude distribution are derived directly from the power spectral density function of equivalent stress or strain, obtained during the reduction of multiaxial loading state to the equivalent uniaxial one. As in the case of cycle counting methods, the fatigue life is determined under an assumption of an adequate hypothesis of damage accumulation.

During literature studies it should be pointed out that in a number of cases it is easier to define load in the frequency domain. It is often necessary to account for the dynamics of the system which affects the loading of its components. In the course of such analyses (conducted e.g. by means of the Finite Element Method (FEM)), power spectral density function of stress or strain are obtained, which could be directly used for the determination of fatigue life by the spectral method. In the contrary case, it is necessary to refer to the time domain, which makes calculation more complex and extends its duration. Therefore the focus in the current book is on the comparison between the cycle counting method and the spectral method by conducting computer simulations devoted to the determination of fatigue life for various stress and strain states and subsequently, a comparison of calculations with experimental data. Chapter 2 is devoted to an overview of bibliography coverage of the spectral methods for determination of fatigue life with a particular focus on multiaxial Gaussian loading.

After the introduction in Chapter 3 of the theoretical essentials of random stress and strain states, the criteria for multiaxial fatigue failure determination in the frequency domain [36, 54, 56] familiar from the literature are stated. Subsequently the assumptions and a technique for the derivation of spectral formulae involving fatigue characteristics $(\sigma_a - N_f)$ and $(\varepsilon_a - N_f)$ are stated. A number of various hypotheses for the fatigue damage accumulation by derivation of fatigue life formula in accordance with Palmgren-Miner, Haibach, Corten-Dolan and Serensen-Kogayev hypotheses are considered.

Chapter 4 is devoted to a description of the algorithm of spectral method for evaluation of fatigue life under multiaxial random loading. A simple calculation model is postulated for determination of the fatigue life of a specific material in a plain stress state. The following Chapter focuses on a description of the simulation along with a comparison between algorithms of determination of fatigue life by means of the cycle counting method and the spectral method. An emphasis in the course of calculations is placed on a comparison between the particular algorithm blocks, such as the computed equivalent values and the determined critical plane positions.

The experimental part of this work – Chapter 6 is devoted to fatigue testing of 18G2A steel with a combination of bending with torsion under random loading. The loading was Gaussian with a narrow-band frequency spectrum. The spectral formula was derived for determination of fatigue life on the basis of the Serenssen-Kogayev linear damage accumulation hypothesis. The computed fatigue life was compared with experimental data. To the verification of fatigue life calculation algorithm applied, experimental data was done by Karolczuk, Niesłony, Achtelik and was conducted as a part of Ph.D. dissertations and papers of Department of Mechanics and Machine Design, Opole University of Technology [47]. In Achtelik's research, the stress history was Gaussian with broad-band frequency spectrum where the specimens under pure bending, pure torsion and proportional bending with torsion were tested.

Spectral Methods for Fatigue Description

2.1 Random Gaussian Loading

Material and structural component fatigue tests require over the test conditions accurate control [10, 41, 60]. One of the methods involves measurement of strain or forces occurring in actual structures during operation and precise representation during laboratory testing. This method is, however, labour and time consuming and is barely applicable with reference to random non-stationary loading. Therefore, real history of loading is adapted by the formation of standard loading, which is postulated for determination of fatigue life of the investigated structural components. Sequences applied in automobile industry: CARLOS [16, 87] and in construction: DABM, TR440, BRE [105] (loading generated by wind in particular climatic zones) could be listed here.

In a number of particular cases it is desirable to present the frequency structure of loading or state a direct definition of loading in frequency domain [24, 25, 30, 63, 65, 70, 97]. The power spectral density (PSD) function of loading is the most commonly applied one. A number of standard PSD functions are postulated, similarly as in time domain. The issues are covered widely by Pook [77], Bitner-Gregersen and Cramer [14] by a comparison of several analytical PSD functions applied for the description of loading generated by the waving sea. However, an awareness of the frequency structure does not give a sufficient description of loading. It is additionally necessary to learn the probability distribution of instantaneous values of loading history. On the basis of observations it was remarked that a number of natural phenomena directly affecting fatigue loading of machines and components are normally distributed. Under this assumption the determination of amplitude distributions in the statistical sense is possible, similar to the distribution gained by means of the cycle counting algorithm [1]. This distribution, along with material constants, constitutes input data for determination of life time with hypotheses of fatigue damage accumulation. An adequate determination of amplitude distribution[1] constitutes a major problem to solve in the spectral method of fatigue life determination.

Bishop and Sherratt [13] submit an algorithm for stress range estimation directly from the power spectral density function. The function of range probability density, similarly as Rychlik [81, 82], is formally defined as a product of the probability of three occurrences P_1, P_2 and P_3 constituting prerequisites of determination of a cycle from random load. A graphical representation of cycle definition is presented in Fig 2.1. k-range cycle probability density is defined as follows:

$$p_k(k) = \frac{2.0}{dk} \sum_{ip=k+1}^{ip=\infty} P_1(ip, ip - k)\, P_2(ip, ip - k)\, P_3(ip, ip - k)\, P(ip), \quad (2.1)$$

where: k – range expressed in terms of the number of classes with a defined width dk,

ip – number which includes the first point of a determined cycle, point 1 in Fig 2.1,

$ip - k$ – class number which includes the second point of a determined cycle, point 2 in Fig 2.1,

P_1 – probability of occurrence that signal reaching point 1 crosses the level of point 2 from below,

P_2 – probability of occurrence that signal approaching point 2 from point 1 following arbitrary trajectory must not exceed levels of point 1 or 2,

P_3 – probability of occurrence that signal approaching point 3, which is placed above level of point 1, from point 2 must not exceed level of point 2,

$P(ip)$ – probability of occurrence of peak on the ip level.

It was indicated that the definition of a loading cycle leads to the establishment of cycle probability density (2.1) by determination of functions of P_1, P_2, P_3 occurrence probability and probability of $P(ip)$ peak occurrence. In order to determine the quantities, it is assumed that the time history is Gaussian with arbitrary band frequency width. Subsequently, a two-dimensional probability density function of adjacent extremes is applied, as postulated by Kowalewski [38] for Gaussian processes

$$P_{min,max}(x_{min}, x_{max}) = \frac{(x_{max} - x_{min})}{4m_0 I^2 \sqrt{2\pi m_0(1 - I^2)}}$$
$$\exp\left[\frac{-x_{min}^2 - x_{max}^2 - 2x_{min}x_{max}(2I^2 - 1)}{8m_0 I^2(1 - I^2)}\right],$$
$$(2.2)$$

---

[1] Amplitude distribution is expressed as a probability density function of amplitude or range for distinguished cycles of a random load history.

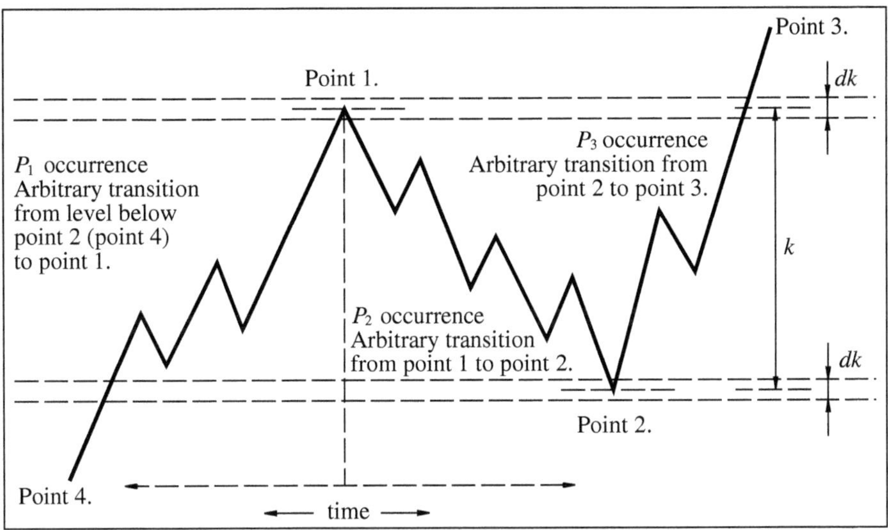

Fig. 2.1. Representation of cycle definition determined from random time history by rain flow algorithm (Bishop and Sherratt [13])

where: x_{min}, x_{max} — local extremes – minimum and maximum,

$$m_k = \int\limits_0^\infty G_{xx}(f)f^k df$$ – k-th moment of power spectral density of process X,

$$I = \frac{m_2}{\sqrt{m_0 m_4}}$$ – irregularity coefficient for process X.

In order to describe the relations between non-adjacent extremes, the Markov chain[2] is applied. On its basis, three transition matrices are developed for the determination of the probability of transition from one level to another, from point 4 to 1, 1 to 2, 2 to 3, respectively (see Fig 2.1).

The derived range probability density function (2.1) is equivalent to the same function determined by direct application of cycle counting by the rain flow algorithm.

Etube et al. [23] postulate the determination of the power spectral density function of stress in the components of a jack-up platform with a transmittance function. In the course of analysis of an offshore platform this was largely simplified by representation as a combination of a cylindrical base and a cube as the platform. By the application of equation of forced vibration motion:

$$M\ddot{x}(t) + C\dot{x}(t) + Kx(t) = F(t)\,, \tag{2.3}$$

---

[2] Application of Markov chain processes for calculation of fatigue strength is widely covered by Sobczyk and Spencer [90] and Johannesson [34].

where: M, C, K – mass, damping, stiffness, respectively,

$\ddot{x}(t), \dot{x}(t), x(t)$ – acceleration, velocity and transition, respectively,

$F(t)$ – time variable force

under an assumption of steady state, i.e. one in which due to damping the transient decayed, a formula[3] for spectral transmittance of the investigated system is derived

$$H(\omega) = \frac{1}{(K - M\omega^2) + j\,(C\omega)}\,, \tag{2.4}$$

which is applied for determination of power spectral density of dislocation

$$G_{xx}(f) = |H(f)|^2\, G_{yy}(f)\,, \tag{2.5}$$

where: $H(f)$ – spectral transmittance in frequency domain,

$G_{xx}(f)$ – power spectral density of dislocation,

$G_{yy}(f)$ – power spectral density of loading (force).

In the course of structure analysis modified analytic function of Pierson-Moskovitz power spectral density [71] is adapted

$$G_{yy}(f) = \frac{H_s^2 T_z}{8\pi^2}\,[(f - \beta)T_z]^{-5} \exp\left\{-\frac{1}{\pi}\,[(f - \beta)T_z]^{-4}\right\}\,, \tag{2.6}$$

where: H_s, T_z – mean wave height and period – values taken from tables dependent on forecasted sea state and platform position,

$(f - \beta)$– frequency diminished by correction coefficient depending on type of platform structure.

On the basis of (2.4) it could be remarked that spectral transmittance takes into account the effect of structure inertia, rigidity and damping and is not dependent on loading. During comparison between function $G_{xx}(f)$ obtained experimentally from measurements of dislocation at several points with functions determined from (2.5), a large degree of equivalence is obtained.

The huge advantage of the application of loading definition in frequency domain is the possibility of application of finite element method for the determination of power spectral density of displacements, strain and stress taking into consideration frequency transmittance of a system [25, 31, 63, 94].

2.1.1 Narrow-Band Frequency Stress

One of earliest papers devoted to spectral methods for determination of fatigue strength is a paper by Miles from 1954 [59]. The author states an assumption that probability density function of peak values of stress time history is equal to amplitude probability density which could be distinguished in random

---

[3] This operation is popular with deriving formulae applied in FEM dynamic analyses [104].

time history. This assumption is only valid for narrow-band frequency histories with the following stages occurring subsequently: local maximum, zero level crossing with negative inclination, local minimum and zero level crossing with positive inclination,[4] followed by further recurrences periodically. By the application of Palmgren-Miner linear hypothesis of damage accumulation and amplitudes approximation by means of Raleigh probability distribution a formula for fatigue life is derived of the form

$$T = \frac{A}{M^+ \, (2m_0)^{\frac{m}{2}} \, \Gamma\left(\dfrac{m+2}{2}\right)}, \qquad (2.7)$$

where: $A = \sigma_a^m N_f$ – coefficient derived from Wöhler curve,

m – Wöhler curve exponent,

$M^+ = \sqrt{\dfrac{m_4}{m_2}}$ – expected number of peaks in a time unit,

$\Gamma(\cdot)$ – gamma function.

The author skilfully adapted the excedance theory [80] developed by Rice for determination of the character and parameters of amplitude probability distribution. It needs to be stated that obtaining such simple relations was possible only under an assumption that instantaneous values of stress history are normally distributed over a narrow frequency band.

The paper by Perruchet and Vimont [69] is devoted to experimental verification of the Miles formula (2.7). The testing involved plain cylindrical specimens made of AU4G1 aluminium alloys applied in aviation components. The tests were conducted on a Schenk 100kN hydraulic machine controlled by force. A random history was reproduced repeatedly from a magnetic tape loop. Throughout the testing the force was recorded and was subsequently subjected to spectral analysis.

Experimental results indicated good correlation with calculated fatigue life. Random loading fatigue life was remarked to spread more than in the case of cyclic loading. In the 1970s the testing conducted by Perruchet and Vimont was considered innovative (random loading, magnetic tape data recording, hydraulic test machine). Currently this research plays historical and cognitive roles.

Another work devoted to experimental verification of formula (2.7) is a paper by Clevenson and Stainer [19]. Investigations involved aluminium alloy 2023-T4, subject to static (tensile test) and dynamic testing for cyclic and random tension-compression. Random loading was stationary Gaussian with zero expected value. Testing included three PSD function forms while the controlled parameter was standard deviation of stress. A solid specimen with circular cross section ($d = 12.7$ mm in diameter) had a grooved notch

---

[4] This type of narrow-band frequency loading is often called variable amplitude loading.

($r = 1.57$ mm in radius), which implied a theoretical stress concentration coefficient of 2.2. Fatigue test machine included an electromagnetic actuator. A
comparison of computed fatigue life with experimental data indicated equivalent results for large stress levels. For small standard deviations of stress, the
calculations indicate exceeded life time, which is attributed to the fallibility
of the applied linear hypothesis of damage accumulation. A specific feature of
the investigated material involves a similarity of the surfaces of fatigue failure
for the case of cyclic and random loading.

The extension of the formula for fatigue life of narrow-band frequency into
cases of loading with two dominant angular frequencies ω_1 and ω_2 is proposed
by Sakai and Okamura [84]. The expected life time to material failure was
determined under an assumption that total damage is obtained by summation
of a component process with dominant angular frequency ω_1 and the other
with dominant angular frequency ω_2 with appropriate weight

$$T = \frac{2^{1-\frac{m}{2}}\pi A}{\Gamma\left(\dfrac{m+2}{2}\right)} \frac{1}{\omega_1 + \omega_2} \left(\omega_1 \frac{m_{0,1}^{1-m}}{m_{2,1}} + \omega_2 \frac{m_{0,2}^{1-m}}{m_{2,2}}\right), \qquad (2.8)$$

where: $m_{0,1}$, $m_{2,1}$ – respective zero and second moment of power spectral density of component process with dominant angular fre-
quency ω_1,

$m_{0,2}$, $m_{2,2}$ – respective zero and second moment of power spectral density of component process with dominant angular fre-
quency ω_2.

Sakai and Okamura indicate that the model correlates well with results
gained by the cycle counting algorithm and damage accumulation, on the
condition that the two dominant angular frequencies ω_1 and ω_2 are sufficiently
distant. The proximity of the positions brings about interference, in which case
the assumption of weight summation of two independent processes is false.

Fu and Cebon [26] indicate that the Sakai and Okamura model does not
account for the amplitude magnification for a lower frequency process. By the
simplification of a random process with two dominant frequencies it could be
represented as a sum of two sinusoid histories with various frequencies. This
case is presented in Fig. 2.2, in which maximum amplitude of the process is
described as the sum of component parts, i.e. $A_{max} = A_1 + A_2$.

This leads to a conclusion that an interaction occurs between component
histories which brings about a necessity of accounting for it in calculations.
Fu and Cebon apply the linear Palmgren-Miner hypothesis for damage accumulation resulting from process $y(t) = y_1(t) + y_2(t)$ with a formula

$$D = \frac{n_1}{N_1} + \frac{n_2 - n_1}{N_2} = \frac{T}{2\pi A}\left[\omega_1 \sigma_{a1}^m + (\omega_2 - \omega_1)\sigma_{a2}^m\right]. \qquad (2.9)$$

$$y(t) = A_1 \sin(\omega_1 t) + A_2 \sin(\omega_2 t)$$

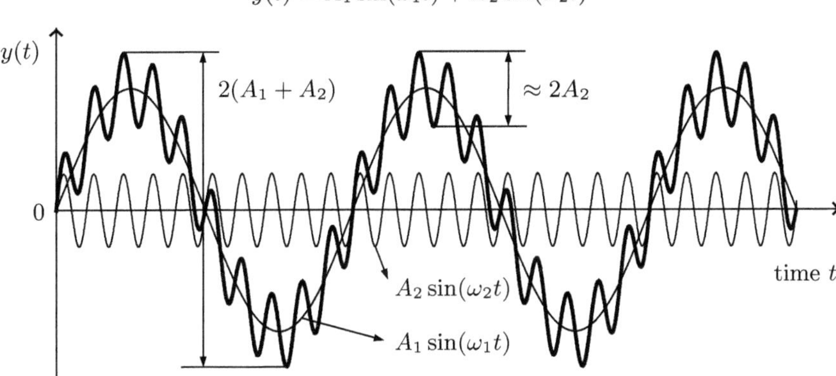

Fig. 2.2. Representation of Fu and Cebon model assumptions [26]

where: n_1 – cycles with amplitude $\sigma_{a1} = A_1 + A_2$,

$n_2 - n_1$ – cycles with amplitude $\sigma_{a2} = A_2$,

N_1, N_2 – cycles derived from Wöhler curve for stress amplitudes σ_{a1} and σ_{a2}, respectively.

By extension into the case in which component processes $y_1(t)$ and $y_2(t)$ have a narrow-band frequency spectra, the formula is derived

$$D = \frac{T}{2\pi A} \left[\omega_1 \int_0^\infty \frac{p_{\sigma_{a1}}(\sigma_a)}{\sigma_a^{-m}} d\sigma_a + (\omega_2 - \omega_1) \int_0^\infty \frac{p_{\sigma_{a2}}(\sigma_a)}{\sigma_a^{-m}} d\sigma_a \right], \qquad (2.10)$$

where the distribution of amplitudes $\sigma_{a1} = A_1 + A_2$ is described by the probability density function

$$p_{\sigma_{a1}}(\sigma_a) = \frac{1}{m_{0,1} m_{0,2}} \exp\left(\frac{-\sigma_a^2}{2m_{0,2}^2} \right)$$

$$\int_0^\infty (\sigma_a y - y^2) \exp\left[\frac{\sigma_a y}{m_{0,2}} - \frac{y^2}{2m_{0,1}} - \frac{y^2}{2m_{0,2}} \right] dy, \qquad (2.11)$$

and the distribution of amplitudes $\sigma_{a2} = A_2$ is approximated by the Rayleigh probability density function.

The final fatigue life formula can be derivered by the transformation of formula (2.10) under the assumption that damage $D = 1$. Difficulties are encountered associated with the separate treatment of component processes $y_1(t)$ and $y_2(t)$ during the application of the formula. It is difficult to explicitly determine the dominant frequencies and moments of the component processes by familiarity with power spectral density of their sums.

A similar issue constitutes the focus of research conducted by Jiao in paper [33]. A theoretical model for the determination of fatigue life is considered in the circumstances of loading under a sum of two processes, a normal one with narrow-band frequency and shock loading. Figure 2.3 represents a section of a loading history in an ideal form. $X_1(t)$ process is substituted with a sinusoidal history, while $X_{U1}(t)$ and $X_{U2}(t)$ are narrow-band frequency processes with dominant frequency higher then from $X_1(t)$ process and damping.

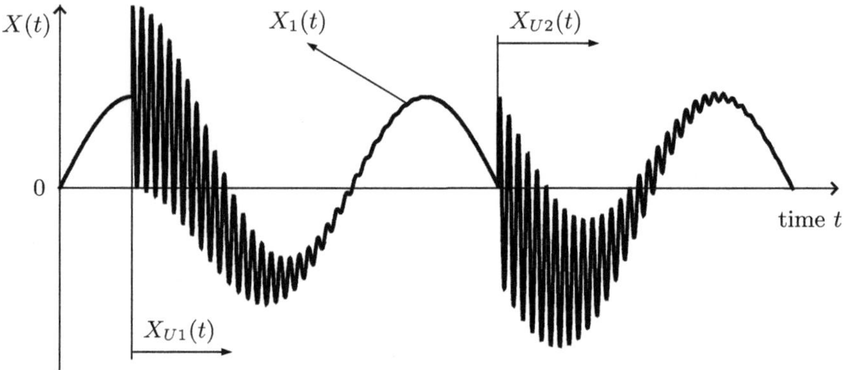

Fig. 2.3. Representation of a section of a loading history conducted by Jiao [33]. $X_1(t)$ history for narrow-band frequency, $X_{U1}(t)$ and $X_{U2}(t)$ – shock faults

Under such assumptions, the following fatigue life formula was determined

$$T = \frac{2^{-m} A T_0 \left[1 - \exp\left(-2\pi\xi m\right)\right]}{\displaystyle\int_0^\infty r_2^m p_{R_2}(r_2) dr_2 + \frac{N_0^+}{\mu_{X_1}} \int_0^{T_0} \int_0^\infty \int_0^\infty (z + \alpha r_2)^m \, z \exp\left(-\frac{z^2}{2\mu_{X_1}}\right) p_{R_2}(r_2) dz dr_2 dt},$$

(2.12)

where: T_0 – mean duration of shock load X_U,

        ξ – damping coefficient,

    $p_{R_2}(r_2)$ – probability density of constituent ranges of shock loading.

A comparison of calculations with life time gained by means of the cycle counting method indicate a large degree of equivalence. Such solutions are applied in the course of fatigue life determination for ships with large displacement. $X_1(t)$ process defines stress resulting from motion while $X_{U1}(t)$ and $X_{U2}(t)$ fault – from sea waves in the direction of the bow. In this case, the application of the spectral formula enables quick fatigue analysis as a result of sufficient knowledge of sea surface behavior [14, 24].

In the course of this spectral formula statement, a hypothesis of fatigue damage accumulation is assumed. The linear Palmgren-Miner hypothesis is applied [12, 59] along with the amplitude cycles while summing fatigue damage. Liou et al. [42] postulate application of Morrow hypothesis[5] [61] for accounting for plastic strain work

$$D = \sum_i \frac{n_i}{N_{f\,i}} \left(\frac{\sigma_{a\,i}}{\sigma_{max}} \right)^{d_M} ,$$

(2.13)

where: σ_{max} – maximum amplitude in analysed stress history,
d_M – exponent including the effect of plastic strain work on fatigue damage.

d_M coefficient describes sensitivity of materials to loading history (amplitude distribution). In accordance with excedance theory in the case of Gaussian history with narrow-band frequency the distribution of amplitudes approaches the Rayleigh distribution. The formula (2.13) could then be restated in the following form

$$\begin{aligned}
D &= N_0^+ T \int_0^\infty \frac{p(\sigma_a)}{N_f(\sigma_a)} \left(\frac{\sigma_a}{\sigma_{max}} \right)^{d_M} d\sigma_a \\
&= \frac{N_0^+ T}{A\mu_\sigma} \int_0^\infty \frac{\sigma_a^{m+d_M+1}}{\sigma_{max}^{d_M}} \exp \left(-\frac{\sigma_a^2}{2\mu_\sigma} \right) d\sigma_a ,
\end{aligned}$$

(2.14)

where: μ_σ – variance of stress $\sigma(t)$.

It needs to be remarked that the integral in the preceding equation is calculated in the limit 0 to ∞, which indicates that damage accumulation involves all the cycles with amplitudes value from 0 to $+\infty$. In practice, however, during testing in a lab, a limit of stress amplitude σ_{max} is determined in order to prevent the fatigue test stand from damage. In this case all of the values $\sigma_a > \sigma_{max}$, which result from normally distributed stress history, are converted into the maximum admissible ones. Upon consideration of the fact, the damage formula takes the form

$$\begin{aligned}
D =& \frac{N_0^+ T}{A\sigma_{max}^{d_M}} (2\mu_\sigma)^{\frac{m+d_M}{2}} \Gamma \left(\frac{m+d_M}{2} + 1, \frac{\sigma_{max}^2}{2\mu_\sigma} \right) \\
&+ \frac{N_0^+ T}{A} \sigma_{max}^{m-d_M} (2\mu_\sigma)^{\frac{d_M}{2}} \left[\Gamma \left(\frac{d_M}{2} + 1 \right) - \Gamma \left(\frac{d_M}{2} + 1, \frac{\sigma_{max}^2}{2\mu_\sigma} \right) \right] ,
\end{aligned}$$

(2.15)

where the first term of the equation represents damage resulting from amplitudes up to the limit value ($\sigma_a < \sigma_{max}$) while the other one represents damage

---

[5] The same hypothesis of fatigue damage accumulation is postulated by Corten and Dolan [20, 95] but under different theoretical assumptions

resulting from maximum amplitude values ($\sigma_a = \sigma_{max}$). Liou et al. indicate that formula (2.15) under the assumptions $d_M = 0$ and $\sigma_{max} \to \infty$ could be simplified, thus deriving the Miles formula (2.7). The analysed paper [42] includes a number of remarks and formulae with practical application, which constitute a basis for the modification of formula (2.15). One of the formulae is

$$N_f = N_0^+ T,\tag{2.16}$$

for determination of damage D in the function of cycle number N_f under an assumption that the analysis involves narrow band frequency history. By the approximation of amplitude by the Rayleigh distribution, the standard deviation could be derived directly from the variance of stress history μ_σ

$$\sqrt{\mu_{\sigma_a}} = \sqrt{\mu_\sigma \left(2 - \frac{\pi}{2}\right)}.\tag{2.17}$$

Following from that, formula (2.15) could be modified into

$$
\begin{aligned}
D = &\frac{N_f}{A\sigma_{max}^{d_M}} \left(\frac{\sqrt{\mu_{\sigma_a}}}{1 - \pi/4}\right)^{m+d_M} \Gamma\left(\frac{m + d_M + 2}{2}, \frac{(1 - \pi/4)\sigma_{max}^2}{\mu_{\sigma_a}}\right) \\
&+ \frac{N_f}{A}\sigma_{max}^{m-d_M} \left(\frac{\sqrt{\mu_{\sigma_a}}}{1 - \pi/4}\right)^{d_M} \\
&\left[\Gamma\left(\frac{d_M + 2}{2}\right) - \Gamma\left(\frac{d_M + 2}{2}, \frac{(1 - \pi/4)\sigma_{max}^2}{\mu_{\sigma_a}}\right)\right],
\end{aligned}
\tag{2.18}
$$

where damage is determined on the basis of material constants A, m, d_M and two parameters associated with loading – amplitude variance μ_{σ_a} and maximum amplitude σ_{max}. It is the most applicable solution, particularly desirable during fatigue testing where the quantities are the controlled parameters. Maximum stress amplitude σ_{max} is commonly a familiar parameter and results from maximum level on fatigue testing machine for safety purposes. If it is unknown, it could be determined in accordance with the asymptotic theory of extreme statistics

$$\sigma_{max} = \sqrt{\frac{\mu_{\sigma_a}}{2 - \frac{\pi}{2}}} \left(\sqrt{2 \ln N_f} + \frac{0.577}{\sqrt{2 \ln N_f}}\right)\tag{2.19}$$

on condition that N_f is sufficiently large.

Experimental verification was performed on aluminium alloy 7075-T651. The conducted fatigue testing in the conditions of constant strain amplitude serve for the determination of fatigue characteristics. Random fatigue testing is conducted under constant standard deviation of stress amplitude with the Rayleigh distribution. The experimentally registered fatigue life was compared with calculations on the basis of Eq. (2.18) and the cycle counting method for various exponents $d_M = 0.0$ (the Palmgren-Miner damage accumulation hypothesis) and $d_M = -0.25; -0.35; -0.45$. The testing indicates similarity of results between calculated fatigue life with experiments under the application of the Morrow hypothesis (2.13) and exponent $d_M = -0.45$. The damage determined by means of the cycle counting method indicated lower level and larger scatter in comparison to proposal (2.18). Although it is not stated directly, the adaptation of the formula (2.18) to the spectral method by the determination of stress history variance $\mu_\sigma = m_0$ from PSD function is possible.

2.1.2 Broad-Band Frequency Stress

The assumption of narrow-band frequency stress history imposes considerable restrictions on the application of the Miles formula due to the fact that majority of random loading have broad-band frequency [30, 77, 90]. A number of proposals for the solution of the problem have been postulated.

Kowalewski [38] accounts for width frequency loading spectrum by an introduction of irregularity coefficient I, which is determined on the basis of power spectral density moments. The following formula for fatigue life is proposed:

$$T = \frac{A}{I^m M^+ (2m_0)^{\frac{m}{2}} \, \Gamma\left(\dfrac{m+2}{2}\right)} \, . \tag{2.20}$$

Rajcher [79] postulates another algorithm of the determination of the expected number of cycles (peaks) in a time unit M^+. His formula for life time calculation takes the form:

$$T = \frac{A}{\left[\displaystyle\int_0^\infty G_0(f) f^{\frac{2}{m}} \, df\right]^{\frac{m}{2}} (2m_0)^{\frac{m}{2}} \, \Gamma\left(\dfrac{m+2}{2}\right)} \, , \tag{2.21}$$

where: $G_0(f) = \dfrac{G(f)}{m_0}$ – normalised power spectral density function of stress.

It could be easily remarked that the Rajcher formula for determination of the expected number of peaks in time unit accounts for coefficient m of the relation describing the Wöhler curve. This leads to the correction of the parameter by accounting for characteristics $(\sigma_a - N_f)$ and results from Rajcher conclusions based on experiment. Below is an suitable transformation of formula for determination of the expected number of zero level transitions in a time unit N_0^+ and Rajcher postulate which reveals their similarity

$$N_0^+ = \sqrt{\frac{m_2}{m_0}} = \sqrt{\frac{\overline{\int_0^\infty G(f)f^2 df}}{\int_0^\infty G(f)df}} = \left[\frac{\int_0^\infty G(f)f^2 df}{\int_0^\infty G(f)df}\right]^{\frac{1}{2}} = \left[\int_0^\infty G_0(f)f^2 df\right]^{\frac{1}{2}},$$

$$\tag{2.22}$$

$$\left[\int_0^\infty G_0(f)f^{\frac{2}{m}} df\right]^{\frac{m}{2}} = \left[\frac{\int_0^\infty G(f)f^{\frac{2}{m}} df}{\int_0^\infty G(f)df}\right]^{\frac{m}{2}}. \tag{2.23}$$

Bolotin [15] states a similar solution postulating the replacement of the expected number of peaks M^+ with the expected number of zero level transitions with a positive slope N_0^+. For narrow-band frequency stress history the fatigue life is identical with the Miles formula (2.7). The Bolotin postulate is additionally less susceptible to the noise included in the analysed stress history due to the application of lower order moments (m_0 and m_2). Similar to Rajcher, Bolotin includes normalised power spectral density function in the final formula

$$T = \frac{A}{\left[\int_0^\infty G_0(f)f^2 df\right]^{\frac{1}{2}} (2m_0)^{\frac{m}{2}} \Gamma\left(\frac{m+2}{2}\right)}. \tag{2.24}$$

Rice postulates the peaks probability density function of Gaussian history with broad-band frequency in his paper [80],

$$p(x) = \frac{1}{\sqrt{2\pi\mu}}\epsilon \exp\left(\frac{-x^2}{2\mu\epsilon^2}\right) + \frac{x}{2\mu}I\left[1 + \mathrm{erf}\left(\frac{x}{\sqrt{\mu(2I^{-2}-2)}}\right)\right]\exp\left(\frac{-x^2}{2\mu}\right),$$

$$\tag{2.25}$$

where: μ – variance of Gaussian random history,
$\epsilon = \sqrt{1-I^2}$ – spectrum width parameter,

later quoted by Sobczyk and Spencer [90].

This distribution is applied by Lü and Jiao [44] and Chow and Li [18] for determination of weighted average amplitude σ_{aw} in accordance with the Palmgren-Miner linear hypothesis of damage accumulation

$$\sigma_{aw} = \sqrt{2\mu}\left[\frac{\epsilon^{m+2}}{2\sqrt{\pi}}\Gamma\left(\frac{m+1}{2}\right) + \frac{I}{2}\Gamma\left(\frac{m+2}{2}\right) + IZ\right]^{\frac{1}{m}}, \tag{2.26}$$

where

$$Z = \int_{0}^{\infty} \mathrm{erf}(x) \left(\frac{\epsilon x}{I}\right)^{m+1} \exp\left[-\left(\frac{\epsilon x}{I}\right)^{2}\right] d\left(\frac{\epsilon x}{I}\right), \qquad (2.27)$$

$$\mathrm{erf}(x) = \frac{2}{\sqrt{\pi}} \int_{0}^{x} \exp(-t^{2}) dt. \qquad (2.28)$$

In order to simplify the calculations, in the further part of Chow and Li paper [18], the transcendental function $\mathrm{erf}(x)$ called error function is replaced with Maclaurin series. A similar approach is adopted by Lü and Jiao [44] by replacement of the error function with two series, which results in gaining better equivalence of results with approximated function $\mathrm{erf}(x)$ than the calculations for the case of Maclaurin series.

The applicability of weighted average amplitude determined in this way for strength calculation[6] is doubted due to the fact that mean values of component cycles are not accounted for. This course is criticised by Dowling [22] as the various cycle counting methods are verified. The papers [18, 44] do not include experimental verification of the postulated solutions.

Wirsching and Light [103] postulate a coefficient λ for correction of fatigue life T derived from formula (2.24)

$$T_{BB} = \frac{T}{\lambda}, \qquad (2.29)$$

to obtain the fatigue life T_{BB} for broad-band frequency loading.

This coefficient constitutes an empirical function related to spectrum width parameter ϵ and the Wöhler curve exponent m

$$\lambda(m, \epsilon) = a(m) + [1 - a(m)](1 - \epsilon)^{b(m)}, \qquad (2.30)$$

where $a(m) = 0.926 - 0.033m$ and $b(m) = 1.587m - 2.323$ are experimentally determined functions.

Determination of the final form of coefficient λ was conducted under following assumptions:

- considerations are devoted to high cycle fatigue regime of material described with relation $(\sigma_a - N_f)$,
- linear damage accumulation in accordance with the Palmgren-Miner hypothesis is assumed,
- stress amplitudes are determined by means of the rain flow algorithm,
- simulation is performed for 4 peculiar power spectral densities.

In the latter part of the paper [103], a sample calculation involves fatigue life determination of the components of a rig. In this case non stationary random loading was described in terms of 11 stationary states accounted for

---

[6] Application of peak distribution of stress could be compared with the peak counting method described in ASTM standard [1].

by analytic functions of power spectral density [11, 14]. Familiar with fraction t_i for each of 11 states fatigue life is determined

$$T_{BB} = \sum_{i=1}^{11} \frac{T_i}{\lambda_i t_i}\,. \tag{2.31}$$

Larsen and Lutes [40, 45] postulate a spectral formula for determination of damage for broad-band frequency loading in the form:

$$D = \frac{2^{\frac{3m}{2}}}{2\pi A}\,\Gamma\left(\frac{m+1}{2}\right) m_{2/m}^{\frac{m}{2}}\,, \tag{2.32}$$

where

$$m_{2/m} = \int_0^\infty f^{\frac{m}{2}} G(f)\,df\,. \tag{2.33}$$

It is called the single moment spectral method due to the characteristic application of single moment of PSD function $m_{2/m}$. In paper [40], a number of simulations are performed for the purpose of making comparison with other formulae for damage determination in accordance with proposals by Miles (2.7), Wirsching and Light (2.29), Oritz and Chen [67]

$$D = \frac{2^{\frac{3m}{2}}}{2\pi A}\,\Gamma\left(\frac{m+1}{2}\right) m_{2/m}^{\frac{m}{2}}\,\frac{m_2^{\frac{m-1}{2}} m_4^{\frac{1}{2}}}{\left(m_{2+2/m}\right)^{\frac{m}{2}}}\,, \tag{2.34}$$

where

$$m_{2+2/m} = \int_0^\infty f^{2+\frac{m}{2}} G(f)\,df\,. \tag{2.35}$$

The analysis involves fatigue life determined on the basis of three various single mode and two double mode analytic PSD functions. In order to compare the calculated fatigue life with life gained by cycle counting method, the stress history is generated directly from PSD function. In a majority of cases more equivalent results are registered by the comparison of the single moment spectral method (2.32) and the cycle counting method in time domain for damage determination than for any other models. An advantage of the single moment spectral method is the simple formula and short calculations.

Chaudhury and Dover [17] remark that the generalised form of probability density distribution of peaks (2.25) could be simplified to the Rayleigh distribution for narrow-band frequency history (while $\epsilon \to 0$) and to normal distribution for broad-band frequency history (while $\epsilon \to 1$). Under an assumption of the linear Palmgren-Miner damage accumulation hypothesis weighted average amplitudes for this two extremes take the following form

$$\sigma_{awNB} = 2\sqrt{2\mu} \left[\Gamma \left(\frac{m+2}{2} \right) \right]^{\frac{1}{m}} , \qquad (2.36)$$

$$\sigma_{awBB} = 2\sqrt{2\mu} \left[\frac{1}{2\sqrt{\pi}} \Gamma \left(\frac{m+1}{2} \right) \right]^{\frac{1}{m}} , \qquad (2.37)$$

where: σ_{awNB}, σ_{awBB} – weighted average amplitudes for narrow- and broad-band frequency stress histories, respectively.

It is indicated that the weighted average amplitude calculated for the distinguished cycles from various stress histories resulting from the schematisation by the rain flow and range counting algorithms is situated between σ_{awNB} and σ_{awBB} extremes. Therefore it is postulated that the calculated fatigue life should be derived from sums of both weighted average amplitudes with the particular weight relative to frequency spectrum width. The weights of particular components are determined on the basis of the Monte Carlo simulation method with an assumption that fatigue life is equal to the life determined by the damage accumulation of cycles determined through the rain flow algorithm

$$T = \frac{A}{\sum\limits_i t_i M_i^+ \left(2\sqrt{2m_{0i}} \right)^m \left[\frac{\epsilon^{m+2}}{2\sqrt{\pi}} \Gamma \left(\frac{m+1}{2} \right) + \frac{3I}{4} \Gamma \left(\frac{m+2}{2} \right) \right]_i} , \qquad (2.38)$$

where: t_i – time fraction for i-th component process,

M_i^+ – expected number of peaks in a time unit for i-th process,

m_{0i} – zero moment (variance) of i-th process.

The general form provides for the determination of fatigue life for Gaussian loading with arbitrary width of the frequency band. Additionally, similar as in [103] if stress history displays non stationary character the authors postulate the separation of the components stationary processes and individual treating them. The final formula then constitutes a sum of the component damages.

Chaudhury and Dover solution [17] is criticised by Tunna [100], who remarks that mean values of component cycles of broad-band frequency stress histories, are not accounted for. A coefficient for compensation for the inconsistency is postulated. It is familiar [52, 48] that in many cases the mean values of component cycles could be ignored in the process of accumulation, if the expected value of total stress history $\sigma_m \approx 0$.

Rychlik [83] proves applicability of the following inequality by assumption of the Gaussian random history in the course of analysis

$$\mathrm{e}\left[D_{RC} \right] \leq \mathrm{e}\left[D_{RFC} \right] \leq \mathrm{e}\left[D_{NB} \right] , \qquad (2.39)$$

where the particular expected damage is determined on the basis of amplitude distribution derived from the range pair algorithm $\mathrm{e}\left[D_{RC} \right]$, rain flow algorithm $\mathrm{e}\left[D_{RFC} \right]$ and approximation by the Rayleigh probability distribution $\mathrm{e}\left[D_{NB} \right]$.

Inequality (2.39) is applied by Benasciutti and Tovo [4, 5] who postulate determination of $\mathrm{e}\,[D_{RFC}]$ as weighted sum of damages $\mathrm{e}\,[D_{NB}]$ end $\mathrm{e}\,[D_{RC}]$

$$\mathrm{e}\,[D_{RFC}] = \mathcal{W}_D \mathrm{e}\,[D_{NB}] + (1 - \mathcal{W}_D)\mathrm{e}\,[D_{RC}] \,. \tag{2.40}$$

Benasciutti and Tovo in papers [4, 5, 99] present two postulations for coefficient $\mathcal{W}_D$ determination

$$\mathcal{W}_D = \min\left(\frac{\alpha_1 - \alpha_2}{1 - \alpha_1}, 1\right), \tag{2.41}$$

$$\mathcal{W}_D = \frac{(\alpha_1 - \alpha_2)\left\{1.112\left[1 + \alpha_1\alpha_2 - (\alpha_1 + \alpha_2)\right]e^{2.11\alpha_2} + (\alpha_1 - \alpha_2)\right\}}{(\alpha_2 - 1)^2}, \tag{2.42}$$

where: $\alpha_k = \dfrac{m_k}{\sqrt{m_0 m_{(2\cdot k)}}}$ for $k = 1, 2$ – coefficient describing the shape of PSD function, for $k = 2$, $\alpha_2 = I$.

The expected damage $\mathrm{e}\,[D_{NB}]$ was determined by transformation of fatigue life formula (2.7) postulated by Miles [59] to the following form

$$\mathrm{e}\,[D_{NB}] = \frac{T_o}{A}M^+ (2m_0)^{\frac{m}{2}}\,\Gamma\left(\frac{m+2}{2}\right), \tag{2.43}$$

while $\mathrm{e}\,[D_{RC}]$ was derived from the Madsen formula [57]

$$\mathrm{e}\,[D_{RC}] \cong I^{m-1}\frac{T_o}{A}M^+ (2m_0)^{\frac{m}{2}}\,\Gamma\left(\frac{m+2}{2}\right) = I^{m-1}\mathrm{e}\,[D_{NB}] \,. \tag{2.44}$$

By the comparison of formula (2.43) with (2.44), the difference between them is visible, which results from the addition of expression I^{m-1} to formula (2.44). This fact is applied for transformation of the formula (2.40) to the simplified form

$$\mathrm{e}\,[D_{RFC}] = \mathrm{e}\,[D_{NB}]\left[\mathcal{W}_D + (1 - \mathcal{W}_D)I^{m-1}\right] \,. \tag{2.45}$$

It could be remarked that determination of damage under loading with the arbitrary width of frequency band is reduced to the determination of damage by the assumption of the Rayleigh amplitude distribution $\mathrm{e}\,[D_{NB}]$, calculations of irregularity coefficient I and adoption of adequate weight coefficient $\mathcal{W}_D$.

The postulated thought Benasciutti and Tovo [4] model of damage determination is verified on the basis of two simulations [5]. The first involves a comparison between the damage calculated under the postulated model (2.45) with two models gained from literature – Wirsching [103] and Dirlik [11, 21], which are often cited and considered as relatively accurate with regard to damage determined by the cycle counting method with application of the rain flow algorithm. In the course of comparative study the Gaussian random loading is generated with irregularity coefficient $I = 0.1$. It is indicated that the largest equivalence of results between the range counting method and postulated formula (2.45) is observed for weight coefficient W_D in accordance with (2.42).

The other simulation involves determination of the fatigue life of a simplified discrete model of a vehicle by consideration of wheels and suspension separately. Roughness of the road is accounted for by modelling it as stochastic process with the known power spectral density function. Finite parameter discrete model enables derivation of the transition function defining the relation between loading and force in wheels and suspension. The functions are applied for determination of the force history and power spectral density function directly on the basis of the applied loading. The calculation of fatigue life presented in charts [5] by means of the spectral method and range counting method indicate large equivalence.

On the basis of the conducted simulations a conclusion that postulated method of fatigue life determination by means of spectral method is the most accurate one could be drawn. The results correlate best with fatigue life gained by cycle damage accumulation determined by the rain flow algorithm. However, as it was developed only recently it was not verified and must be approached very cautiously. It offers better results in comparison to the Dirlik postulate [21]; however, the probability distribution of amplitude postulated by Dirlik provide for the application of other damage accumulation hypotheses in the course of fatigue life determination, which is not possible under Benasciutti and Tovo postulations.

The issue of structure life time under wind loading is introduced by Holmes [32]. He postulates determination of a simple formula accounting for the mean wind velocity $\hat{U}$. The parameter is related to standard deviation of stress in the investigated structural components

$$\sqrt{\mu_\sigma} = K\hat{U}^n \,, \tag{2.46}$$

where: K – coefficient determined on the basis of static structure analysis,
   n – exponent relative to resonance response of a system.

The mean wind velocity $\hat{U}$ constitutes a random variable. Its probability density is described with the Weibull distribution as

$$p_{\hat{U}}(\hat{U}) = \frac{\varrho\hat{U}^{\varrho-1}}{c^\varrho} \exp\left[-\left(\frac{\hat{U}}{c}\right)^\varrho\right] \,, \tag{2.47}$$

where: ϱ – shape coefficient,
   c – scale coefficient.

The application of the distribution for the approximation of real time data it is remarked that values of shape coefficient ϱ are approximately equal to 2. In this case the Weibull distribution assumes a specific shape of the Rayleigh distribution. By the assumption of this distribution of mean wind velocity, the lower and higher limits T_{lower} and T_{higher} of the expected fatigue lifes are determined. The lower limit is determined for narrow-band frequency loading and higher – for broad-band loading

$$T_{lower} = \frac{A}{N_0^+ \left(\sqrt{2}K\right)^m c^{mn} \Gamma\left(\frac{m}{2}+1\right) \Gamma\left(\frac{mn+k}{k}\right)}, \qquad (2.48)$$

$$T_{upper} = \frac{T_{lower}}{\lambda}, \qquad (2.49)$$

where: λ – empirical coefficient accounting for broad-band frequency loading, following [103].

It is remarkable that the above formulae account for probability distribution parameters of the mean wind velocity $\hat{U}$, which accelerates simulation of structures under various service conditions. Holmes in paper [32] does not target the issue of the effect of mean stress on fatigue life. The postulated solution accounting for wind direction as the weighted sum of damage from all directions is under scrutiny

$$D = \sum_{i=1}^{N} P\left(\theta_i\right) D_i, \qquad (2.50)$$

where: $P\left(\theta_i\right)$ – probability of mean wind from i-th direction.

Totally different assumptions are stated by Tovo [98] who postulates, similar as Nagode and Fajdiga [64], the approximation of cycles determined by the rain flow schematisation with multimodal Weibull distribution

$$p(\Delta\sigma) = \sum_{i=1}^{k} p_i(\Delta\sigma) = \sum_{i=1}^{k} w_i \left(\frac{\beta_i}{\vartheta_i}\right) \left(\frac{\Delta\sigma}{\vartheta_i}\right)^{\beta_i-1} \exp\left[-\left(\frac{\Delta\sigma}{\vartheta_i}\right)^{\beta_i}\right], \qquad (2.51)$$

where: k – number of component distributions named characteristic forms,

w_i – i-th weighted coefficient,

β_i, ϑ_i – shape and scale coefficients for i-th characteristic form.

It is assumed that damage generated by a single cycle with a given range $\Delta\sigma$ constitutes a parameter for the determination of accuracy of cycle distribution approximation. It is remarked that the postulated distribution matches well enough for $k = 1$. Nagode and Fajdiga gain similar results for component distributions $k = 4$. The numerical solution of a system of equations is required for determination of coefficients β_1 and ϑ_1

$$\begin{cases} \vartheta_1 \dfrac{\Gamma\left(\frac{n+1}{\beta_1}+1\right)}{\Gamma\left(\frac{n}{\beta_1}+1\right)} = \dfrac{\sum\limits_{i=1}^{S} N_i \Delta\sigma_i^{n+1}}{\sum\limits_{i=1}^{S} N_i \Delta\sigma_i^{n}} \\[4ex] \vartheta_1^2 \dfrac{\Gamma\left(\frac{n+2}{\beta_1}+1\right)}{\Gamma\left(\frac{n}{\beta_1}+1\right)} = \dfrac{\sum\limits_{i=1}^{S} N_i \Delta\sigma_i^{n+2}}{\sum\limits_{i=1}^{S} N_i \Delta\sigma_i^{n}} \end{cases} , \qquad (2.52)$$

where: S – number of stress range levels $\Delta\sigma$,
       N_i – number of cycles with stress range $\Delta\sigma_i$,
$n = 3$ – constant.

Weighted coefficient w_1 in this case is equal to

$$w_1 = \frac{1}{N_{Tot}} \frac{\sum\limits_{i=1}^{S} N_i \Delta\sigma_i^{n}}{\vartheta_1^n \Gamma\left(\frac{n}{\beta_1}+1\right)}, \qquad (2.53)$$

where: N_{Tot} – sum of cycles determined by means of the rain flow algorithm
            from random history.

The approximation of the amplitude or range distribution from random loading constitutes a common technique of the formal recording and definition of parameters of fatigue life determination algorithms. A valuable sample of this approach is represented by papers by Sobczykiewicz [91] and Oziemski [68]. Service histories of excavator component loading are registered and analysed. A linear combination of two theoretical probability distributions, exponent distribution and extreme distribution for description of stress amplitude distribution is postulated.

Szala [95] put together a compilation of formulae of fatigue life calculation with the assumption of various amplitude distribution and damage accumulation hypotheses in a book devoted to fatigue damage accumulation hypotheses. The total number of the programmed load cycles to fatigue failure is determined for Rayleigh and exponent distributions and right leg normal distribution in the combination with the Palmgren-Miner and Haibach linear damage accumulation hypotheses. The form of loading spectrum frequency is accounted for by means of an adequate spectrum with parameter.

Unfortunately, the discussed postulations by Tovo, Nagode, Sobczykiewicz, Oziemski and Szala find no direct application with regard to spectral methods despite good approximation of amplitude probability distribution. The problem of the direct determination of parameters from stress history and range counting in probability distributions is encountered. The parameter cannot be determined if the input data only include power spectral density function.

Dirlik [21] postulates the approximation of amplitude distribution gained by rain flow algorithm by means of an empirical formula derived from the Monte Carlo simulation method. It is assumed that the zero moment m_0 of stress power spectral density function constitutes the most relevant parameter. On the basis of 17 representative stress histories and their power spectral density functions the standard error between the assumed model and real amplitude distribution is minimised. The derived probability density function of ranges of stress cycles takes the form

$$p(\Delta\sigma) = \frac{1}{2\sqrt{m_0}} \left[\frac{G_1}{Q} e^{\frac{-z}{Q}} + \frac{G_2 Z}{R^2} e^{\frac{-z^2}{2R^2}} + G_3 Z e^{\frac{-z^2}{2}} \right], \qquad (2.54)$$

where: $Z = \dfrac{\Delta\sigma}{2\sqrt{m_0}},$ $G_1 = \dfrac{2(x_m - I^2)}{1 + I^2},$

$G_2 = \dfrac{1 - I - G_1 + G_1^2}{1 - R},$ $G_3 = 1 - G_1 - G_2,$

$R = \dfrac{I - x_m - G_1^2}{1 - I - G_1 + G_1^2},$ $Q = \dfrac{1.25(I - G_3 + G_2 R)}{G_1},$

$x_m = \dfrac{m_1}{m_0} \left(\dfrac{m_2}{m_4} \right)^{\frac{1}{2}},$ $I = \dfrac{m_2}{\sqrt{m_0 m_4}}.$

According to authors [11, 12, 60, 62] more accurate amplitude distribution approximation by rain flow algorithm [1, 81] is gained by application of this formula in comparison to other approximations, and it can be applied for stress histories with narrow- and broad-band frequency spectra. Unfortunately, the complicated form of the function does not enable derivation of a simple life time formula

$$T = \frac{1}{M^+ \displaystyle\int_0^\infty \frac{p(\Delta\sigma)}{N(\Delta\sigma)} d\Delta\sigma} \qquad (2.55)$$

and the above integral must be calculated numerically.

Morrill et al. [60] apply probability distribution developed by Dirlik (2.54) for the procedure of the determination of characteristic PSD function applied in fatigue testing machines. The paper undertakes the issue of precise mapping of service conditions during fatigue testing with the concurrent assumption of their short duration. The decisive parameter during determination of characteristic PSD functions for a specific loading type is damage calculated on the basis of amplitude probability distribution (2.54) and the adoption of the Palmgren-Miner hypothesis of damage accumulation. The functions are determined on the basis of histories of acceleration registered with car suspension components during road tests in various conditions. The following relation is applied

$$G_\sigma\left(f_i\right) = \frac{C}{f_i^4} G_{\ddot{x}}\left(f_i\right) , \tag{2.56}$$

where: $G_\sigma\left(f_i\right)$ – discrete stress PSD function value for f_i frequency,
$\quad\quad G_{\ddot{x}}\left(f_i\right)$ – discrete acceleration PSD function value for f_i frequency,
$\quad\quad C$ $\quad\quad$ – scalar constant,

in order to derive power spectral density of stress from spectral density of acceleration. The authors postulate determination of constant C under the finite element method or calibration method by measurement of acceleration and strain in the investigated structural points. In the particular cases the analytical determination of constant C would be difficult.

The registered acceleration histories subjected to the analysis are non-stationary. By the statistical analysis of history sections, stationary sections are distinguished and divided into groups with regard to statistical similarity. Mean acceleration and variance variation in time are analysed. PSD function is determined for the distinguished sections. Weighted mean values are calculated for individual groups

$$\hat{G}(f) = \sum_{i=1}^{k} \left[\frac{G_i(f)T_i}{T_\Sigma}\right] , \tag{2.57}$$

where: $T_\Sigma = \sum_{i=1}^{k} T_i$ – total time for a single group,

$\quad\quad \hat{G}(f)$ $\quad\quad\quad$ – mean PSD function for single group,
$\quad\quad G_i(f)$ $\quad\quad\quad$ – PSD function for i-th section,
$\quad\quad k$ $\quad\quad\quad\quad$ – total number of stationary sections distinguished within a single group,
$\quad\quad T_i$ $\quad\quad\quad\quad$ – length of i-th section,

which results in a characteristic PSD function for a group, $\hat{G}(f)$. After determination of power spectral density of stress $G_\sigma(f)$ on the basis of (2.56) damage is determined on the basis of Dirlik postulate (2.54). The determined damage and fundamental statistical parameters for distinguished stationary history groups constitute the foundation for distinguishing groups for fatigue lab testing. For the currently analysed example loading history the method contributes to the reduction of testing duration from 1300 to 20 hours. In order to trace the complex process of determination of specific PSD the following algorithm steps are conducted:

- registration of loading (acceleration history),
- selection of data block length in seconds for statistical analysis,
- statistical analysis of particular blocks,
- definition of a finite group number with various statistical parameters,
- attribution of blocks with similar statistical parameters into appropriate groups,

- determination of weighted mean PSD function in accordance with (2.57),
- conversion of weighted mean PSD acceleration function into stress PSD function in accordance with (2.56),
- determination of PSD moments and damage by means of the Dirlik formula for each group,
- selection of distinctive groups taking damage, mean value and variance of stress.

2.2 Random Non-Gaussian Loads

The literature devoted to random non-Gaussian loads in spectral methods is scarce. One of the first ones is Liu and Hu paper [43], which is devoted to the adaptation of damage D_{NB} determined with the narrow-band frequency formula (2.7) by accounting for coefficients λ and L. By reference to Wirsching and Light paper [103], the parameter λ (2.30) is applied in order to revise the recent definition of damage by accounting for frequency band width. Parameter L is introduced in order to account for deviation from Gaussian load history. It is determined in the course of simulations by the Monte Carlo method under following assumption

$$L(\kappa, m) = \frac{D\left[Y(t), m\right]}{D\left[X(t, \kappa), m\right]},\tag{2.58}$$

where: $D[\cdot]$ – damage determined by means of the rain flow algorithm and the linear Palmgren-Miner hypothesis,

$Y(t)$ – stress history with non-Gaussian probability distribution,

$X(t, \kappa)$ – Gaussian stress history determined by non-linear transformation of $Y(t)$.

Finally, non-Gaussian loading damage with broad-band frequency spectrum is derived from the formula

$$D = \lambda L D_{NB}.\tag{2.59}$$

Sarkani et al. [85, 86] remark that a number of random natural processes are non-Gaussian and adopt a standard fourth central moment parameter named kurtosis κ_σ for deviation from normal probability distribution history (Fig. 2.4)

$$\kappa_\sigma = \frac{e\left[(\sigma - \hat{\sigma})^4\right]}{\mu_\sigma^2},\tag{2.60}$$

where: $e[\cdot]$ – expected value of an expression,

$\sigma, \hat{\sigma}$ – stress and mean stress, respectively,

μ_σ – variance of the stress.

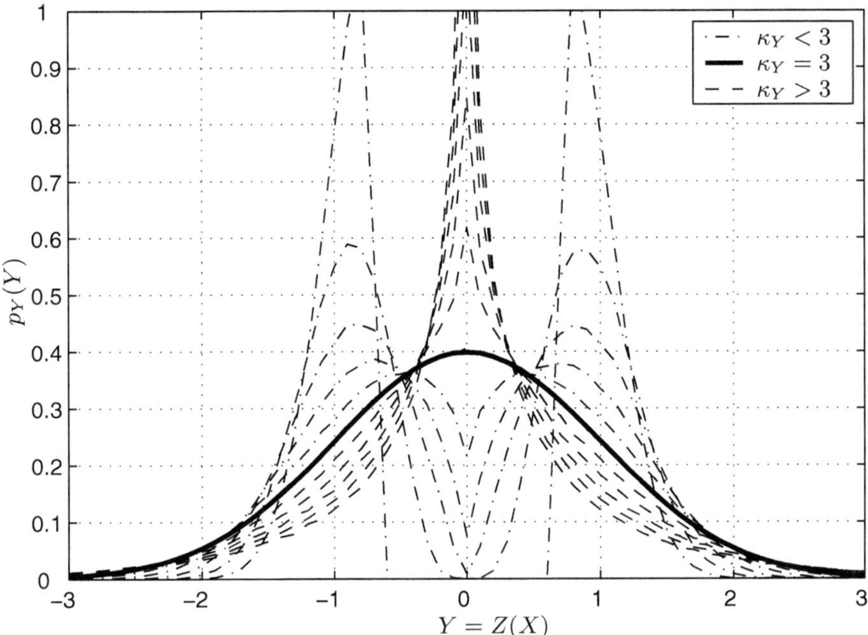

Fig. 2.4. Variation of probability density function shape for various non-Gaussian histories $Y(t)$. Loading histories are derived by means of transformation function Z (2.62). Coefficients β and n are incorporated in function $\beta(n) = n + 2$ for $n = (0; 0.2; 0.4; ...; 2)$ [85, 86]

In general, non-Gaussian history could be determined by means of non-linear transformation of Gaussian history

$$Y(t) = Z\left[X(t)\right], \tag{2.61}$$

where: $Y(t)$ – non-Gaussian history,

$\quad\quad X(t)$ – Gaussian history,

$\quad\quad Z[\cdot]$ – non-linear transformation function.

A number of non-linear functions $Z[\cdot]$ could be postulated for performing the transformation. Sarkani et al. [85, 86] postulated it in the form:

$$Y = Z(X) = \frac{X + \beta\left[\mathrm{sgn}(X)|X|^n\right]}{C}, \tag{2.62}$$

which includes coefficients β and n to account for deviation from Gaussian history. Constant C is introduced to normalise history Y and ensure mean square values of Y and Gaussian history X to be equal. By means of the linear Palmgren-Miner hypothesis, a damage formula for non-Gaussian history is derived for an input kurtosis κ_σ, for which parameters β, n and C are determined

$$D_{ng} = \int\limits_{0}^{\infty} \frac{\sigma_a \left(\sigma_a + \beta\sigma_a^n\right)^m}{C^m A \mu_\sigma} \exp\left(\frac{-\sigma_a^2}{2\mu_\sigma}\right) d\sigma_a \,, \qquad (2.63)$$

where: $A = \sigma_a^m N$ – fatigue characteristic,

μ_σ – variance of the stress history.

A comparison between the postulated theoretical solution and an experiment was made. A welded joint under alternating bending loading was tested for fatigue. The histories applied for investigation were determined by means of transformation function (2.62) for three selected kurtosis values $\kappa_\sigma = 3$, 2 and 5. Four tests were performed for each of five various stress levels. The comparison of calculations with experiment proves its accuracy and indicates good applicability of the postulated calculation model for non-Gaussian loads with narrow-band frequency spectrum.

Similar to Liu and Hu [43], Winterstein [24, 102] postulates application of a coefficient for narrow-band frequency histories whose distribution is different than Gaussian

$$L_W = 1 + \frac{m\left(m - 1\right)\left(\kappa_\sigma - 3\right)}{24} \qquad (2.64)$$

for correction of damage D_{NB} determined by means of narrow-band frequency formula (2.7). It needs to be remarked that coefficient L_W takes the value 1 for Gaussian processes ($\kappa_\sigma = 3$). The final fatigue life formula takes the form

$$T = \frac{A}{M^{+}\left(2m_0\right)^{\frac{m}{2}}\, \Gamma\left(\dfrac{m + 2}{2}\right)\left[1 + \dfrac{m\left(m - 1\right)\left(\kappa_\sigma - 3\right)}{24}\right]}\,. \qquad (2.65)$$

Lachowicz et al. [39] tested 10HNAP steel under broad-band frequency loading and non-Gaussian distribution. Fatigue life calculation is performed in time domain by application of the Palmgren-Miner hypothesis with coefficient $a_{PM} = 0.5$, which takes account of amplitudes below fatigue limit

$$D(T_o) = \sum_{i=1}^{k}\begin{cases} \dfrac{n_i}{N_G \left(\dfrac{\sigma_{af}}{\sigma_{ai}}\right)^m} & \text{for} \quad \sigma_{ai} \geq a_{PM}\sigma_{af} \\[2em] 0 & \text{for} \quad \sigma_{ai} < a_{PM}\sigma_{af} \end{cases}, \qquad (2.66)$$

where: $D(T_o)$ – damage determined for observation time T_o,

n_i – cycle number with amplitude σ_{ai},

σ_{af} – fatigue limit.

Component amplitudes are determined by the rain flow algorithm. Concurrently, calculations by means of spectral method in frequency domain are performed upon the selection of three formulae from literature: Rajcher, Miles,

Kowalewski. Due to non-Gaussian loading, a coefficient L_L is applied for correction of this deviation, similar as in Liu and Hu [43]. The coefficient is relative to kurtosis and is defined as follows

$$L_L = \frac{1}{|1 + (\kappa_\sigma - 3)|}.\tag{2.67}$$

In the course of comparison of experimental data with calculation ones the best equivalence of results was obtained for rain flow algorithm and Rajcher fatigue life formula (2.21) including a correction coefficient L_L.

2.3 Multiaxial Loading

The literature devoted to application of spectral methods with reference to multiaxial random loading is scarce.

Kam and Dover [35] tackle the problem in the course of developing research methodology for pipe joints, typical of drilling rig structures. The rigs applied for crude oil mining on the shallow sea, are subjected to changeable loading resulting from wind and sea waves. The loading was thoroughly investigated and is most commonly described in terms of a number of power spectral density empirical functions [11, 14, 17, 32], whose shape is relative to weather and wavy motion of sea surface. A generation of loading history is developed for the purpose of fatigue testing. The Markov chain is applied for the description of transitions from a single characteristic state to another[7]. The generator operates in a loop performing the following tasks:

- determination of power spectral density of stress for a particular characteristic state; subsequent characteristic states are combined into Markov chain,
- calculation of transmittance function for each of the states,
- calculation of filter function,
- random signal generation by withe noise filtration with normal probability distribution.

A T-joint consisting of a major carrying pipe and an arm of smaller diameter was tested. The arm with the pipe forms right angle. The carrying pipe is fixed at both ends to the testing stand while loading is applied by means of three hydraulic cylinders whose ends are attached to the arm of the tested joint. The loading along centers of the arm with bending in two directions constitute the combined fatigue loading. The actual measurements of structure indicated considerable differences between this three service loadings (PSD function shapes were different), although they where considerably correlated.

---

[7] The issue of application of the Markov chain for separation of a non-stationary process into a series of stationary subprocesses in order to count cycles is covered by Ph.D. thesis by Johannesson [34].

The histories of loads with such properties are gained by the appropriate selection of phase shift of component harmonics during generation. Unfortunately, the paper neither includes the results of fatigue life nor a postulate or proposal of component life time calculations.

Leser et al. [41] postulate another solution of the problem of multiaxial loading modeling for fatigue testing. It is assumed that loading history constitutes a superposition of a stationary random process with the zero expected value and random variables affecting variance and mean value of the modeled process

$$\mathbf{x} = \hat{\mathbf{x}} + \mathbf{s}\bar{\mathbf{x}}, \tag{2.68}$$

where: $\mathbf{x} = [x_1, \ldots, x_n]^T$ – modeled n-dimensional random process,
$\hat{\mathbf{x}} = [\hat{x}_1, \ldots, \hat{x}_n]^T$ – random variable affecting mean value $\mathbf{x}$,

$$\mathbf{s} = \begin{bmatrix} s_{11} & \cdots & s_{1n} \\ \vdots & \ddots & \vdots \\ s_{n1} & \cdots & s_{nn} \end{bmatrix}$$ – symmetrical matrix of $n \times n$ dimension for scaling $\mathbf{x}$ with regard to variance of particular dimension,

$\bar{\mathbf{x}} = [\bar{x}_1, \ldots, \bar{x}_n]^T$ – n-dimensional random stationary process with the zero expected value.

In order to minimise the number of the required parameters for deterministic description of mean value vector $\hat{\mathbf{x}}$ and matrix $\mathbf{s}$ the functions are represented as finite Fourier series. It is possible due to the small variability of the quantities in comparison to the dynamics of random variable $\bar{\mathbf{x}}$ for which modelling ARMA (Auto Regressive Moving Average) model is applied [89]. With reference to the spectral methods Leser et al. remark that the awareness of PSD function for components of multidimensional random variable $\bar{\mathbf{x}}$ and coefficients from the finite Fourier series ($\hat{\mathbf{x}}$ and $\mathbf{s}$ modelling) is sufficient for the adequate mapping of a non-stationary process.

An algorithm of fatigue life determination under multiaxial random loading is first postulated by Macha [56]. It constitutes an extension of the postulates of Miles [59], Kowalewski [38], Rajcher [79] and Bollotin [15]. It consists in the application of power spectral density of the equivalent stress for determination of statistical parameters involved in the familiar fatigue life formulae. PSD for the equivalent stress is determined by the application of the linear criteria of multiaxial fatigue failure in frequency domain. As the criteria are based on the notion of a critical plane, three methods of determination of its location are distinguished:

- weight functions method [54, 55],
- variance method [9],
- damage accumulation method [51, 53].

Despite the fact that none of the above methods is presented from the spectral point of view, a possibility is open for the instance of two latter ones. The paper by Macha [56] could be considered typically theoretical, as it is distinguished with a detailed presentation of the part devoted to the stress criteria of multiaxial random fatigue and their definition in frequency domain. It is observable that Macha maintains the possibility of applying the method for fatigue life calculations by means of strain criteria of multiaxial fatigue failure.

Preumont and Piéfort [78] introduce a method for fatigue life determination under plain stress state by the application of equivalent stress in accordance with the Huber-Mises-Hencky strength hypothesis

$$\sigma_{eq}^2 = \sigma_{xx}^2 + \sigma_{yy}^2 - \sigma_{xx}\sigma_{yy} + 3\sigma_{xy}^2 \,, \tag{2.69}$$

where: σ_{xx}, σ_{yy} and σ_{xy} – components of stress tensor.

After defining vector $\boldsymbol{\sigma} = (\sigma_{xx}, \sigma_{yy}, \sigma_{xy})^T$, the formula (2.69) could be restated in accordance with the principles of matrix calculations

$$\sigma_{eq}^2 = \boldsymbol{\sigma}^T \mathbf{Q} \boldsymbol{\sigma} = \text{Trace}\left\{\mathbf{Q}[\boldsymbol{\sigma}\boldsymbol{\sigma}^T]\right\} \,, \tag{2.70}$$

where: $\mathbf{Q} = \begin{bmatrix} 1 & -0.5 & 0 \\ -0.5 & 1 & 0 \\ 0 & 0 & 3 \end{bmatrix}$ – matrix of coefficients for Huber-Mises-Hencky hypothesis under plane stress state,

$\boldsymbol{\sigma}^T$ – vector transposed to $\boldsymbol{\sigma}$,

$\text{Trace}\{\cdot\}$ – sum of components of main diagonal of square matrix.

On the basis of formula (2.70) the relation for expected values could be stated

$$e\left[\sigma_{eq}^2\right] = \text{Trace}\left\{\mathbf{Q}e\left[\boldsymbol{\sigma}\boldsymbol{\sigma}^T\right]\right\} \,. \tag{2.71}$$

Finally a formula for mean square value of equivalent stress is derived. Mean square value could be determined directly from power spectral density of equivalent stress

$$e\left[\sigma_{eq}^2\right] = \int_0^\infty G_{eq}(f)df = \int_0^\infty \text{Trace}\left\{\mathbf{Q}\mathbf{G}_{\sigma\sigma}(f)\right\}df \,, \tag{2.72}$$

where: $\mathbf{G}_{\sigma\sigma}(f)$ – matrix of autospectral and cross-spectral density functions of stress vector $\boldsymbol{\sigma}$.

On the basis of the preceding formulae the authors postulate a method for determination of power spectral density function of the equivalent stress directly from a matrix of spectral density function of stress vector $\boldsymbol{\sigma}$

$$G_{eq}(f) = \text{Trace}\left\{\mathbf{Q}\mathbf{G}_{\sigma\sigma}(f)\right\} = \sum_{i,j} Q_{ij}G_{\sigma_i\sigma_j}(f) \,, \tag{2.73}$$

where: Q_{ij} – ij indexed coefficient of $\mathbf{Q}$ matrix,

$G_{\sigma_i\sigma_j}(f)$ – power autospectral density ($i = j$) or cross-spectral density ($i \neq j$) functions of vector $\boldsymbol{\sigma}$ components.

It must be remarked that some values of coefficient matrix $\mathbf{Q}$ are equal to zero. It is remarked that in the course of determination of power spectral density function for equivalent stress (2.69), on the basis of formula (2.73) in paper [78], the interaction between the components of stress σ_{xx} and σ_{xy}, and also σ_{yy} and σ_{xy} is not taken into account, which constitutes a fault of the method.

An extension of the above approach is presented by Potoiset et al. [72, 73, 75, 76] in the statement of power spectral density function of the equivalent stress by the application of Matake and Crossland criteria. The criteria are based on concept of a critical plane. Matake makes an assumption that a critical plane is one for which the shear stress amplitude $\tau_{a\eta}$ is the largest. The amplitude is defined as the radius of smallest circle circumscribing vector for the shear stress $\tau_{a\eta}(t)$ in a plane with normal unit vector $\bar{\eta}$ (see Fig. 2.5).

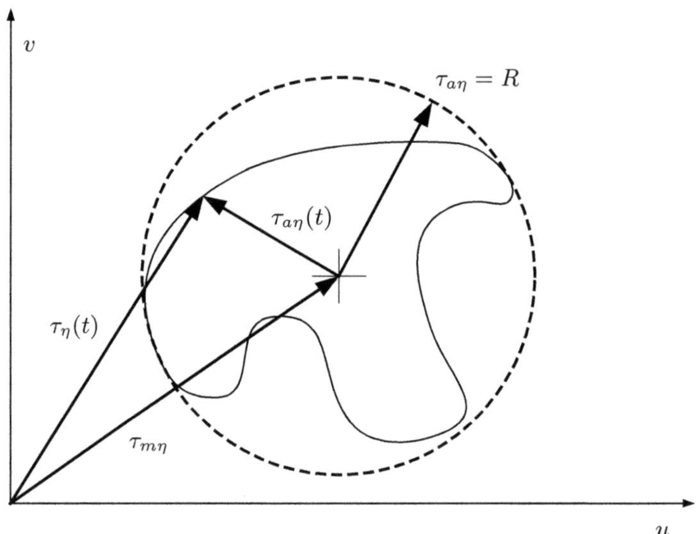

Fig. 2.5. Determination of radius R of circle circumscribing vector for shear stress $\tau_{a\eta}(t)$ on a plane with normal unit vector $\bar{\eta}$ (based on [75])

The criterion based on the assumption takes the form

$$\frac{\tau_{a\eta} + \left(\dfrac{2\sigma_{af}}{\tau_{af}} - 1\right)(\sigma_{m\eta} + \sigma_{a\eta})}{\tau_{af}} \leq 1, \qquad (2.74)$$

where: $\tau_{a\eta}$ – maximum shear stress amplitude in critical plane,
    σ_{af}, τ_{af} – fatigue limit for tension-compression and torsion, respectively,
    $\sigma_{m\eta}$, $\sigma_{a\eta}$ – mean value and amplitude of normal stress to critical plane, respectively.

Preumont et al. solution is applied in a work by Sun end Wang [94] for determination of fatigue life by means of finite element method under plane stress state. The novelty involves the application of a linearized plate component for determination of stress state components. The linear shape was determined under an assumption that modal response of the analysed structure is Gaussian. Unfortunately, the work does not include an experimental verification of the postulated model.

Grzelak et al. [28] conduct spectral analysis of selected multiaxial fatigue failure criteria. Simulations involve the generation of full Gaussian stress tensor components with broad-band frequency spectrum (maximum frequency $f_{max_{ij}} = 160\,\text{Hz}$) and subsequently the application of strength criteria and comparison of statistical parameters of equivalent stress history. Nine criteria of multiaxial fatigue failure are analysed:

- criterion of maximum normal stress in a critical plane,
- criterion of maximum normal strain in a critical plane,
- criterion of maximum shear stress in a critical plane,
- criterion of maximum principal stress (Galileo hypothesis),
- criterion of largest normal stress (Ranky hypothesis),
- criterion of largest normal strain (Saint-Venant hypothesis),
- criterion of largest shear stress (Coulomb-Tresca-Guest hypothesis),
- Huber-Mises-Hencky hypothesis,
- Beltrame hypothesis.

The following conclusions are postulated:

1. Mathematical model of the multiaxial fatigue failure criterion stated as a linear combination of stress tensor components allows to determine statistical parameters applicable during the estimation of life time,
2. Equivalent stresses gained by the application of linear criteria have the same probability distribution function as stress tensor components applied during simulations,
3. Stress tensor frequency band is retained in equivalent stress history only for the case of the linear criteria.

The paper demonstrates that the determination of power spectral density function of the equivalent stress for the case of linear multiaxial fatigue failure criteria is possible directly form the matrix of autospectral and cross-spectral density functions of stress tensor components.

Similar conclusions are drawn by Łagoda and Macha [46]. The analysis focuses on frequency variations of equivalent histories gained by the application of various multiaxial fatigue failure criteria. It is remarked that only for the case of linear criteria the frequency characteristics of equivalent stresses are not altered.

3

Theoretical Fundamentals

3.1 Description of Random Stress and Strain States

As mechanical loadings are observed it is remarked that they are often combined of random quantities. Examples include wind pressure exerted on slender building structures, hitting waves against obstacles, seismic vibrations, etc. In order to formulate a mathematical model of such phenomena, the theory of random functions is applied. Random functions including time parameter t, which are named stochastic processes, are common. A set of a number n of processes constitutes a vectorial, n-dimensional stochastic process

$$\mathbf{X}(t) = [X_1(t), X_2(t), \ldots, X_n(t)] . \tag{3.1}$$

Under the general notion of loading, it induces random state of stress and strain in a structural component. It could be described in terms of stress tensor $\boldsymbol{\sigma}(t)$ and strain tensor $\boldsymbol{\varepsilon}(t)$, respectively, variable in time:

$$\boldsymbol{\sigma}(t) = \begin{bmatrix} \sigma_{xx}(t) & \sigma_{xy}(t) & \sigma_{xz}(t) \\ \sigma_{yx}(t) & \sigma_{yy}(t) & \sigma_{yz}(t) \\ \sigma_{zx}(t) & \sigma_{zy}(t) & \sigma_{zz}(t) \end{bmatrix} , \tag{3.2}$$

$$\boldsymbol{\varepsilon}(t) = \begin{bmatrix} \varepsilon_{xx}(t) & \varepsilon_{xy}(t) & \varepsilon_{xz}(t) \\ \varepsilon_{yx}(t) & \varepsilon_{yy}(t) & \varepsilon_{yz}(t) \\ \varepsilon_{zx}(t) & \varepsilon_{zy}(t) & \varepsilon_{zz}(t) \end{bmatrix} . \tag{3.3}$$

Under the assumption of material isotropy, stress tensor and strain tensor are symmetrical matrices, in which

$$\sigma_{xy}(t) = \sigma_{yx}(t), \qquad \sigma_{xz}(t) = \sigma_{zx}(t), \qquad \sigma_{zy}(t) = \sigma_{yz}(t), \tag{3.4}$$

$$\varepsilon_{xy}(t) = \varepsilon_{yx}(t), \qquad \varepsilon_{xz}(t) = \varepsilon_{zx}(t), \qquad \varepsilon_{zy}(t) = \varepsilon_{yz}(t) . \tag{3.5}$$

It is remarked that under the above assumptions tensors (3.2) and (3.3) could be described with application of their six components. Hence, six-dimensional vectorial stochastic processes for stress and strain states can be derived in the form

$$X_k(t) = \sigma_{ij}(t)\,, \qquad\qquad (k = 1,\ldots,6;\ i,j = x,y,z)\,, \qquad (3.6)$$

$$X_k(t) = \varepsilon_{ij}(t)\,, \qquad\qquad (k = 1,\ldots,6;\ i,j = x,y,z)\,, \qquad (3.7)$$

or

$$\boldsymbol{\sigma}(t) = [\sigma_{xx}(t),\ \sigma_{yy}(t),\ \sigma_{zz}(t),\ \sigma_{xy}(t),\ \sigma_{xz}(t),\ \sigma_{yz}(t)]\,, \qquad (3.8)$$

$$\boldsymbol{\varepsilon}(t) = [\varepsilon_{xx}(t),\ \varepsilon_{yy}(t),\ \varepsilon_{zz}(t),\ \varepsilon_{xy}(t),\ \varepsilon_{xz}(t),\ \varepsilon_{yz}(t)]\,. \qquad (3.9)$$

They describe stress and strain in materials on the basis of specific component tensors related to the Cartesian co-ordinate system $Oxyz$.

In many applications random processes, whose probabilistic characteristics are not variable under time axis shift, are encountered. Such phenomena are modeled by means of so-called stationary stochastic processes. A stochastic process $\mathbf{X}(t)$, $t \in T$, is strongly stationary or stationary in a strict sense if all possible statistical distributions of component processes are independent of an arbitrary time shift τ, provided that $t_i + \tau \in T$. For the current instance, the relation of equality

$$p_{t_1, t_2, \ldots, t_n}\,(x_1,\, x_2,\, \ldots,\, x_n) = p_{t_1+\tau,\, t_2+\tau,\, \ldots,\, t_n+\tau}\,(x_1,\, x_2,\, \ldots,\, x_n, \tau) \quad (3.10)$$

must be satisfied.

In practice it is difficult to verify if a physical process is stationary in a strict sense, because the relation of equality (3.10) must be satisfied for each component vector process and arbitrary time shift τ. Therefore, a wider class of stationary processes is introduced for simplification purposes. A stochastic process $\mathbf{X}(t)$, $t \in T$, is said to be weakly stationary or stationary in a wide sense if its mean (expected) values are time invariant and all elements of correlation matrix $\mathbf{R}(t_1, t_2)$ depend only on the time difference $\tau = t_2 - t_1$, i.e. $\mathbf{R}(t_1, t_1 + \tau) = \mathbf{R}(\tau)$. From the definition it follows that in order to consider a vectorial stochastic process stationary, the fulfilment of the condition of stationary component content is not sufficient. Apart from that, the component processes must be stationary mutually correlated.

If a random process $\mathbf{X}(t)$ is stationary and mean values $\hat{X}_k$ and autocorrelation functions $R_{X_k}(\tau)$ of component processes X_k have equal statistical properties, a multidimensional random process is named ergodic. For an ergodic random process, the mean value and autocorrelation function of single history are equal to the appropriate mean values in a set of histories, so $\hat{X}_k = \hat{X}$ and $R_{X_k}(\tau) = R_X(\tau)$. It must be remarked that only stationary processes can be ergodic. In practice, random processes representing stationary physical phenomena are commonly ergodic. Due to this, in a majority of instances, characteristics of a stationary random process could be sufficiently determined on the basis of a single realisation. In the monograph it was assumed that vectors of stress (3.8) and strain (3.9) are stationary and ergodic processes.

In the correlation theory [6, 7], a stationary and ergodic vectorial process (3.1) is described by means of a vector of expected values $\hat{\mathbf{x}} = [\hat{x}_1,\ \hat{x}_2,\ \ldots,\ \hat{x}_6]$

and a matrix of correlation function $\mathbf{R_x}(\tau)$ or a covariance function $\boldsymbol{\mu_x}(\tau)$, where $\tau = t_2 - t_1$.

Under the assumption that a vectorial process $\mathbf{X}(t)$ is a six-dimensional stationary and ergodic Gaussian process the joint probability density function takes the form

$$p_{x_1,\ldots,x_6}\left(x_1,\ldots,x_6,\tau\right) = \frac{1}{\sqrt{(2\pi)^6 \left|\boldsymbol{\mu_x}(\tau)\right|}} \exp\left[-\frac{1}{2}\mathbf{x}\,\boldsymbol{\mu_x}^{-1}(\tau)\,\mathbf{x}^T\right], \quad (3.11)$$

where: $\boldsymbol{\mu_x}(\tau) = \begin{bmatrix} \mu_{x_{11}}(\tau) & \cdots & \mu_{x_{16}}(\tau) \\ \vdots & \ddots & \vdots \\ \mu_{x_{61}}(\tau) & \cdots & \mu_{x_{66}}(\tau) \end{bmatrix}$ – covariance matrix of random variables $X_1,\ldots,X_6$,

$\mathbf{x} = [x_1 - \hat{x}_1,\ldots,x_6 - \hat{x}_6]$ – row vector for variables $x_1,\ldots,x_6$ and expected values $\hat{x}_1,\ldots,\hat{x}_6$,

$\mathbf{x}^T$ – $\mathbf{x}$ transposed vector,

$\left|\boldsymbol{\mu_x}(\tau)\right|$ – determinant of covariance matrix.

The awareness of the distribution of probability density of instantaneous values and stationary character of random history does not give sufficient information about a stochastic process. However, for the case of stochastic processes there is an instrument of spectral analysis parallel to harmonic analyses of deterministic functions. The analysis is based on the power spectral density function and offers a possibility of gaining additional information about frequency structure of a stochastic process.

For one-dimensional stochastic process X a two-sided power spectral density function $S_X(f)$ is defined as Fourier transform of autocorrelation function[1]

$$S_X(f) = \int\limits_{-\infty}^{\infty} R_X(\tau)e^{-j2\pi f\tau}\,d\tau, \quad (3.12)$$

where: $S_X(f)$ – two-sided power spectral density function defined for frequency range $(-\infty, +\infty)$,

$R_X(\tau)$ – autocorrelation function.

Due to the difficulties in the interpretation of negative frequency, one-sided power spectral density function $G_X(f)$ is applied, in which the argument f varies within a limit $(0, +\infty)$

$$G_X(f) = \begin{cases} 2S_X(f) \text{ for } 0 \le f < \infty, \\ 0 \quad \text{ for } f < 0, \end{cases} \quad (3.13)$$

However, in mathematical calculations the application of function $S_X(f)$, defined for the range $(-\infty, +\infty)$ with an imaginary exponent in an integrand,

---

[1] By replacement of exponent of Euler constant e in Fourier transform from $(-j2\pi f\tau)$ into $(-j\omega\tau)$, two-sided power spectral density defined in angular frequency domain $S_X(\omega)$ is obtained, where $S_X(f) = 2\pi S_X(\omega)$.

simplifies the analysis. It is important to apply both function forms appropriately.

For the investigated example of a six-dimensional stochastic process (3.8) or (3.9), one-sided power spectral density function takes the form of the Hermitian matrix $\mathbf{G}(f)$, 6×6 of dimension

$$\mathbf{G}(f) = \begin{bmatrix} G_{11}(f) & \cdots & G_{16}(f) \\ \vdots & \ddots & \vdots \\ G_{61}(f) & \cdots & G_{66}(f) \end{bmatrix}. \tag{3.14}$$

The functions $G_{kl}(f)$ are defined for frequency $f \geq 0$ and are equal to the double value of two-sided power spectral density $S_{kl}(f)$

$$G_{kl}(f) = \begin{cases} 2S_{kl}(f) & \text{for} \quad 0 \leq f < \infty, \\ 0 & \text{for} \quad f < 0, \end{cases} \quad (k, l = 1, \ldots, 6), \tag{3.15}$$

where: $G_{kk}(f)$, $S_{kk}(f)$ – autospectral density functions of component processes $X_k(t)$,

$G_{kl}(f)$, $S_{kl}(f)$ – cross-spectral density functions between component processes $X_k(t)$ and $X_l(t)$.

Taking into consideration the fact that functions of power spectral density are complex functions

$$G_{kl}(f) = \mathrm{Re}[G_{kl}(f)] + i\,\mathrm{Im}[G_{kl}(f)], \tag{3.16}$$

where: $\mathrm{Re}[G_{kl}(f)]$ – coincident spectral density function, a real part of $G_{kl}(f)$,

$\mathrm{Im}[G_{kl}(f)]$ – quadrature spectral density function, an imaginary part of $G_{kl}(f)$,

$i = \sqrt{-1}$ – imaginary unit.

Since for the Hermitian matrices the relation $G_{kl}(f) = \mathrm{Re}[G_{lk}(f)] - i\,\mathrm{Im}[G_{lk}(f)]$ is fulfilled, in order to characterise frequency structure of a random stress or strain tensors the knowledge of 21 power spectral density functions is required.

3.2 Multiaxial Fatigue Failure Criteria

Numerous criteria of multiaxial fatigue failure introduced in [53, 54] are based on an assumption that material damage is induced by a combination of stress and strain components operating on the critical plane. On the basis of the assumptions, general function of material fatigue strength under multiaxial random loading could be defined

$$S(t) = \{D_{ij}(t),\, P_n,\, C_k\}, \tag{3.17}$$

where: $D_{ij}(t)$ – components of stress or strain tensor; stochastic processes,
$\quad\quad P_n$ – parameters for determination of critical plane position,
$\quad\quad C_k$ – parameters characterizing a material.

The surface of limit state determining fatigue life under random multiaxial state of stress (or strain) is described by the maximum value of strength function

$$\max_t \{S(t)\}\,, \tag{3.18}$$

which is equivalent to fatigue strength of material under uniaxial cyclic loading. On the basis of [9, 53, 54], two general multiaxial fatigue failure criteria could be distinguished:

1. Generalized criterion of maximum normal and shear stresses in the critical plane for long life time,
2. Generalized criterion of maximum normal and shear strains in the critical plane for ling and short life times.

Assumptions for the two criteria can be written as:

1. Fatigue damage depends directly on normal stress $\sigma_\eta(t)$ (normal strain $\varepsilon_\eta(t)$) and shear stress $\tau_{\eta s}(t)$ (shear strain $\varepsilon_{\eta s}(t)$) in direction $\bar{s}$ in the critical plane with normal vector $\bar{\eta}$;
2. Direction $\bar{s}$ is consistent with the mean direction of maximum shear stress $\max_s \{\tau_{\eta s}(t)\}$ (maximum shear strain $\max_s \{\varepsilon_{\eta s}(t)\}$) in the critical plane;
3. In the limit state the maximum value of strength function (3.18), which is a linear combination of stresses $\sigma_\eta(t)$ and $\tau_{\eta s}(t)$ (strains $\varepsilon_\eta(t)$ and $\varepsilon_{\eta s}(t)$) satisfies the following equations:
   – in accordance with stress criterion

$$\max_t \{B\tau_{\eta s}(t) + K\sigma_\eta(t)\} = F\,, \tag{3.19}$$

   – in accordance with strain criterion

$$\max_t \{b\varepsilon_{\eta s}(t) + k\varepsilon_\eta(t)\} = q\,, \tag{3.20}$$

where constants B and b are applied for the selection of particular criterion form. Constants K, F, k and q refer to fatigue characteristics of materials and are gained from uniaxial cyclic tests. The positions of unit vectors $\bar{\eta}$ and $\bar{s}$ are determined from mean directional cosines of principal stress or principal strain axes l_n, m_n, n_n, $(n = 1, 2, 3)$.

At the time of the analysis of experimental results under multiaxial cyclic and random loading it was observed that for the case of brittle materials the fracture plane is perpendicular to normal stress with the highest amplitude or variance. For plastic materials, the fracture plane takes one of two positions for which shear stresses reach maximum amplitude or variance [55]. By selection of adequate constants B, K, F and b, k, q and specifying the critical plane position, the particular forms of stress (3.19) and strain criteria (3.20) are derived. The following multiaxial fatigue failure criteria could be distinguished and give the resulting formulae of equivalent damage parameters, i.e. equivalent stress and strain [53, 54].

1. *Criterion of maximum normal stress in the critical plane,*
   $(B = 0,\ K = 1)$ – the critical plane is determined by the mean position of the maximum principal stress $\sigma_1(t)$

$$\sigma_{eq}(t) = l_1^2\sigma_{xx}(t) + m_1^2\sigma_{yy}(t) + n_1^2\sigma_{zz}(t)$$
$$+ 2l_1m_1\sigma_{xy}(t) + 2l_1n_1\sigma_{xz}(t) + 2m_1n_1\sigma_{yz}(t)\,. \tag{3.21}$$

2. *Criterion of maximum shear stress in the critical plane,*
   $(B = 1,\ K = 0)$ – the critical plane is determined by the mean position of one of two planes in which the maximum shear stress $\tau_1(t)$ occurs

$$\sigma_{eq}(t) = \left(l_1^2 - l_3^2\right)\sigma_{xx}(t) + \left(m_1^2 - m_3^2\right)\sigma_{yy}(t) + \left(n_1^2 - n_3^2\right)\sigma_{zz}(t)$$
$$+ 2\left(l_1m_1 - l_3m_3\right)\sigma_{xy}(t) + 2\left(l_1n_1 - l_3n_3\right)\sigma_{xz}(t) \tag{3.22}$$
$$+ 2\left(m_1n_1 - m_3n_3\right)\sigma_{yz}(t)\,.$$

3. *Criterion of maximum normal and shear stresses in the critical plane,*
   $(B = 1,\ K \neq 0)$ – the critical plane is determined by the mean position of one of two planes in which the maximum shear stress $\tau_1(t)$ acts

$$\sigma_{eq}(t) = \frac{1}{1+K}\left\{ \left[l_1^2 - l_3^2 + K\left(l_1^2 + l_3^2\right)^2\right]\sigma_{xx}(t)\right.$$
$$+ \left[m_1^2 - m_3^2 + K\left(m_1^2 + m_3^2\right)^2\right]\sigma_{yy}(t)$$
$$+ \left[n_1^2 - n_3^2 + K\left(n_1^2 + n_3^2\right)^2\right]\sigma_{zz}(t) \tag{3.23}$$
$$+ 2\left[l_1m_1 - l_3m_3 + K\left(l_1 + l_3\right)\left(m_1 + m_3\right)\right]\sigma_{xy}(t)$$
$$+ 2\left[l_1n_1 - l_3n_3 + K\left(l_1 + l_3\right)\left(n_1 + n_3\right)\right]\sigma_{xz}(t)$$
$$\left.+ 2\left[m_1n_1 - m_3n_3 + K\left(m_1 + m_3\right)\left(n_1 + n_3\right)\right]\sigma_{yz}(t)\right\}\,.$$

4. *Criterion of maximum normal strain in the critical plane,*
   $(b = 0,\ k = 1)$ – the critical plane is determined by the mean position of the maximum principal strain $\varepsilon_1(t)$

$$\varepsilon_{eq}(t) = l_1^2\varepsilon_{xx}(t) + m_1^2\varepsilon_{yy}(t) + n_1^2\varepsilon_{zz}(t)$$
$$+ 2l_1m_1\varepsilon_{xy}(t) + 2l_1n_1\varepsilon_{xz}(t) + 2m_1n_1\varepsilon_{yz}(t)\,. \tag{3.24}$$

5. *Criterion of maximum shear strain in the critical plane,*
   $(b = 1,\ k = 0)$ – the critical plane is determined by the mean position of one of two planes in which the maximum shear strain $\gamma_1(t)$ occurs

$$\varepsilon_{eq}(t) = \frac{1}{1+\nu} \left[\left(l_1^2 - l_3^2 \right) \varepsilon_{xx}(t) + \left(m_1^2 - m_3^2 \right) \varepsilon_{yy}(t) + \left(n_1^2 - n_3^2 \right) \varepsilon_{zz}(t) \right.$$
$$+ 2 \left(l_1 m_1 - l_3 m_3 \right) \varepsilon_{xy}(t) + 2 \left(l_1 n_1 - l_3 n_3 \right) \varepsilon_{xz}(t)$$
$$\left. + 2 \left(m_1 n_1 - m_3 n_3 \right) \varepsilon_{yz}(t) \right] ,$$

$$(3.25)$$

where: ν – Poisson's ratio.

6. *Criterion of maximum normal and shear strains in the critical plane,*
   ($b = 1$, $k = 1$) – the critical plane is determined by the mean position of
   one of two planes in which the maximum shear strain $\gamma_1(t)$ acts

$$\varepsilon_{eq}(t) = l_1 \left(l_1 + l_3 \right) \varepsilon_{xx}(t) + m_1 \left(m_1 + m_3 \right) \varepsilon_{yy}(t) + n_1 \left(n_1 + n_3 \right) \varepsilon_{zz}(t)$$
$$+ \left[l_1 \left(2m_1 + m_3 \right) + l_3 m_1 \right] \varepsilon_{xy}(t) + \left[l_1 \left(2n_1 + n_3 \right) + l_3 n_1 \right] \varepsilon_{xz}(t)$$
$$+ \left[m_1 \left(2n_1 + n_3 \right) + m_3 n_1 \right] \varepsilon_{yz}(t) .$$

$$(3.26)$$

The linear form of the quoted multiaxial fatigue failure criteria enable
presentation of equivalent stress and strain in a more generalised form. Row
vector of coefficients is defined as

$$\mathbf{a} = [a_1, \ldots, a_6] . \qquad (3.27)$$

Equivalent history could be determined by the sum of the products of
suitable component tensors $x_k(t)$ and coefficients a_k dependent on the selected
criterion and the position of the critical plane

$$x_{eq}(t) = \sum_{k=1}^{6} a_k x_k(t) , \qquad (3.28)$$

where: a_k — suitable criterion dependent coefficients (Table 3.1),
$x_k(t)$ – components of random stress or strain tensor.

The conclusion that equivalent history has the same type of probability
density function as the stochastic process (tensor) from which it was determined results from the above linear formula. In the case of Gaussian vector
process $\mathbf{X}(t)$ (3.11), the equivalent history has normal distribution expressed
with probability density function in the form:

$$p_{x_{eq}} \left(x_{eq} \right) = \frac{1}{\sqrt{2\pi \mu_{x_{eq}}}} \exp \left[\frac{- \left(x_{eq} - \hat{x}_{eq} \right)^2}{2\mu_{x_{eq}}} \right] , \qquad (3.29)$$

where the expected value of the process is derived as follows

$$\hat{x}_{eq} = \sum_{k=1}^{6} a_k \hat{x}_k \qquad (3.30)$$

Table 3.1. Components of coefficient vector **a** for three particular stress and strain related criteria.

Stress criteria			
a	(3.21)	(3.22)	(3.23)
a_1	l_1^2	$(l_1^2 - l_3^2)$	$\frac{1}{1+K}\left[l_1^2 - l_3^2 + K\left(l_1^2 + l_3^2\right)^2\right]$
a_2	m_1^2	$(m_1^2 - m_3^2)$	$\frac{1}{1+K}\left[m_1^2 - m_3^2 + K\left(m_1^2 + m_3^2\right)^2\right]$
a_3	n_1^2	$(n_1^2 - n_3^2)$	$\frac{1}{1+K}\left[n_1^2 - n_3^2 + K\left(n_1^2 + n_3^2\right)^2\right]$
a_4	$2l_1 m_1$	$2\left(l_1 m_1 - l_3 m_3\right)$	$\frac{2}{1+K}\left[l_1 m_1 - l_3 m_3 + K\left(l_1 + l_3\right)\left(m_1 + m_3\right)\right]$
a_5	$2l_1 n_1$	$2\left(l_1 n_1 - l_3 n_3\right)$	$\frac{2}{1+K}\left[l_1 n_1 - l_3 n_3 + K\left(l_1 + l_3\right)\left(n_1 + n_3\right)\right]$
a_6	$2m_1 n_1$	$2\left(m_1 n_1 - m_3 n_3\right)$	$\frac{2}{1+K}\left[m_1 n_1 - m_3 n_3 + K\left(m_1 + m_3\right)\left(n_1 + n_3\right)\right]$
Strain criteria			
a	(3.24)	(3.25)	(3.26)
a_1	l_1^2	$\frac{1}{1+\nu}\left(l_1^2 - l_3^2\right)$	$l_1\left(l_1 + l_3\right)$
a_2	m_1^2	$\frac{1}{1+\nu}\left(m_1^2 - m_3^2\right)$	$m_1\left(m_1 + m_3\right)$
a_3	n_1^2	$\frac{1}{1+\nu}\left(n_1^2 - n_3^2\right)$	$n_1\left(n_1 + n_3\right)$
a_4	$2l_1 m_1$	$\frac{1}{1+\nu}\left(l_1 m_1 - l_3 m_3\right)$	$l_1\left(2m_1 + m_3\right) + l_3 m_1$
a_5	$2l_1 n_1$	$\frac{1}{1+\nu}\left(l_1 n_1 - l_3 n_3\right)$	$l_1\left(2n_1 + n_3\right) + l_3 n_1$
a_6	$2m_1 n_1$	$\frac{1}{1+\nu}\left(m_1 n_1 - m_3 n_3\right)$	$m_1\left(2n_1 + n_3\right) + m_3 n_1$

and variance

$$\mu_{x_{eq}} = \sum_{k=1}^{6}\sum_{l=1}^{6} a_k a_l \mu_{x_{kl}} = \sum_{k=1}^{6}\left(a_k^2 \mu_{x_{kk}} + 2\sum_{l=1}^{k-1} a_k a_l \mu_{x_{kl}}\right). \tag{3.31}$$

It results from formula (3.30) that if the expected values of the components of process $\mathbf{X}(t)$ are equal to zero ($\hat{x}_k = 0$; $k = 1, \ldots, 6$), the expected value of the equivalent process $\hat{x}_{eq} = 0$. It must be remarked that the occurrence of zero expected value in the equivalent history with zero expected values of components of process $\mathbf{X}(t)$ constitutes an important result from the physical point of view. The result is not obtained if multiaxial loading is reduced to an uniaxial one with the application of non-linear fatigue failure criteria (e.g. Huber-Mises-Hencky or Tresca strength hypotheses) [28].

3.3 Power Spectral Density Function of Equivalent History

Frequency structure of the equivalent history of the damage parameter, which is defined as a function of power spectral density $G(f)$ [6, 7], plays an important role during calculations of fatigue life. For the case of multiaxial loading, power spectral density of the equivalent stress or strain $G_{x_{eq}}(f)$ should be determined on the basis of suitable criteria of fatigue failure. The simplest technique consists in determination of PSD from the equivalent stress or strain history $x_{eq}(t)$ by means of a numerical method. The objective of the current chapter is to derive it directly from power spectral density matrices (3.14) by the application of the theory of linear multi-input systems [6, 7].

In the investigated example, the equivalent stress or strain $x_{eq}(t)$ could be interpreted as output signal from a six input physical system (see Fig. 3.1). The system has six input signals $x_k(t)$ representing suitable tensor components $\sigma_{ij}(t)$ or $\varepsilon_{ij}(t)$.

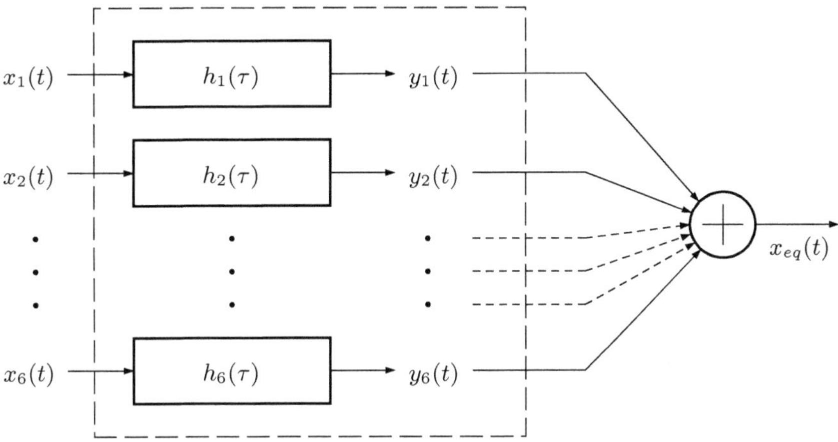

Fig. 3.1. Interpretation of damage parameter $x_{eq}(t)$ as output signal from a physical system with impulse transfer functions $h_k(\tau)$, $k = 1, \ldots, 6$ with input feed of signals $x_k(t)$, $k = 1, \ldots, 6$

For the investigated system the output signal $x_{eq}(t)$ is derived under formula

$$x_{eq}(t) = \sum_{k=1}^{6} \int_{0}^{\infty} h_k(\tau) x_k(t - \tau) d\tau, \tag{3.32}$$

and its autocorrelation function

$$R_{x_{eq}}(\tau) = e\left[x_{eq}(t) x_{eq}(t + \tau)\right]. \tag{3.33}$$

Power spectral density function $G_{x_{eq}}(f)$ is determined with Fourier transform of autocorrelation function $R_{x_{eq}}(\tau)$ [6, 7]

$$G_{x_{eq}}(f) = \sum_{k=1}^{6} \sum_{l=1}^{6} H_k^*(f) H_l(f) G_{kl}(f), \tag{3.34}$$

where: $H_l(f)$ – spectral transmittance function for l-th input (Fourier transform of impulse transfer function $h_l(\tau)$),

$H_k^*(f)$ – $H_k(f)$ coupled function.

For the investigated criteria of multiaxial fatigue failure, the spectral transmittance of the system $H_k(f)$ for the particular input signals do not depend on frequency f and are equal to the constant coefficients, i.e. $H_k(f) = a_k$. Hence, power spectral density of the equivalent history $G_{x_{eq}}(f)$, under an assumption that random components of process $\mathbf{X}(t)$ are correlated, could be derived from the formula

$$
\begin{aligned}
G_{x_{eq}}(f) &= \sum_{k=1}^{6}\sum_{l=1}^{6} a_k a_l G_{kl}(f) \\
&= \sum_{k=1}^{6} a_k^2 G_{kk}(f) + 2\sum_{k<l}^{6} a_k a_l \mathrm{Re}\left[G_{kl}(f)\right] .
\end{aligned}
\tag{3.35}
$$

The above formula enables determination of power spectral density function of the equivalent history directly from the components of the power spectral density matrix (3.14) by the reduction of the spatial state to the equivalent uniaxial state in frequency domain [28]. It could be remarked that the form of the formula (3.35) is similar to formula (3.31) of equivalent history variance. It could have been anticipated, as PSD function reflects mean square values of the harmonic components of the signal.

The formerly derived formulae could be restated more briefly by means of a matrix notation, which enables result presentation in a more comprehensible form. By the application of coefficient row vector $\mathbf{a}$ (3.27), covariance matrix $\boldsymbol{\mu}_{\mathbf{x}}$, and PSD matrix $\mathbf{G}_{\mathbf{x}}(f)$ (3.14) the formulae (3.31) and (3.35) can be presented in the form

$$
\mu_{x_{eq}} = \mathbf{a}\boldsymbol{\mu}_{\mathbf{x}}\mathbf{a}^T ,
\tag{3.36}
$$

$$
G_{x_{eq}}(f) = \mathbf{a}\mathbf{G}_{\mathbf{x}}(f)\mathbf{a}^T .
\tag{3.37}
$$

It should be remarked that in the course of determination of power spectral density function of the equivalent history, auto- and cross-PSD function of matrix $\mathbf{G}_{\mathbf{x}}(f)$, which is a complex matrix, is applied. As a consequence of the summation (3.37), which results from matrix operations, the imaginary parts of matrix component functions $G_{kl}(f)$ and $G_{lk}(f)$ are subtracted and the result takes the form of PSD of the equivalent history $G_{x_{eq}}(f)$ in real number domain. This is a property of Hermitian matrices [37] as illustrated in the example.

Example: For a Hermitian matrix $\mathbf{X}$ (2×2 of dimension) and vector $\mathbf{b}$ (1×2 of dimension), matrix operations according to (3.37) are performed

$$
Z = \mathbf{b}\mathbf{X}\mathbf{b}^T .
$$

Matrix $\mathbf{X}$ and vector $\mathbf{b}$ are defined as follows

$$\mathbf{X} = \begin{bmatrix} x_{11} & x_{12} \\ x_{21} & x_{22} \end{bmatrix},$$

$$\mathbf{b} = \begin{bmatrix} b_1 & b_2 \end{bmatrix}$$

Consequently the form is obtained

$$Z = \mathbf{b}\mathbf{X}\mathbf{b}^T = \begin{bmatrix} b_1 & b_2 \end{bmatrix} \begin{bmatrix} x_{11} & x_{12} \\ x_{21} & x_{22} \end{bmatrix} \begin{bmatrix} b_1 \\ b_2 \end{bmatrix} =$$

$$\begin{bmatrix} b_1 x_{11} + b_2 x_{21} & b_1 x_{12} + b_2 x_{22} \end{bmatrix} \begin{bmatrix} b_1 \\ b_2 \end{bmatrix} =$$

$$(b_1 x_{11} + b_2 x_{21}) b_1 + (b_1 x_{12} + b_2 x_{22}) b_2 =$$

$$b_1^2 x_{11} + b_1 b_2 x_{21} + b_2 b_1 x_{12} + b_2^2 x_{22} =$$

$$b_1^2 x_{11} + b_2^2 x_{22} + b_1 b_2 (x_{21} + x_{12}).$$

The sum $(x_{21} + x_{12})$ from the preceding formula is the sum of components of Hermitian matrix for which the relation $ix_{kl} = -ix_{lk}$ is satisfied. The imaginary parts are reduced, which provides for the restatement of the presented example of equation as:

$$Z = \mathbf{b}\mathbf{X}\mathbf{b}^T = b_1^2 x_{11} + b_2^2 x_{22} + 2 b_1 b_2 \mathrm{Re}\,(x_{12}).$$

3.4 Amplitude Distribution in Spectral Methods

Service loading histories of machines are most commonly complicated random histories. Under such circumstances the application of cycle counting methods for the determination of material life time leads to the schematisation of random loading. The procedure of schematisation offers a possibility of counting amplitudes and mean values of cycles from random time history. Subsequently, in the process of damage accumulation, the cycles are considered to result from cyclic loading. The sequence-bound structure of the schematisation of random histories makes it a considerable share of computational time. For the reason, suitable theoretical probability distributions are often applied for the description of loading peaks (local extremes), mean values and amplitudes distributions [13, 91, 95]. The most important factor from the point of view of material life time is distribution of peaks, which is associated with the *issue of excedance* familiar from theory of random processes.

The distribution of peaks is determined here for a one-dimensional stationary process represented with stress history $\sigma(t)$ with expected value $\hat{\sigma} = 0$ and fulfilling the condition $d\sigma(t)/dt > 0$. The expected value of frequency of level crossing N_a^+ is computed for level $x = a$ where $a > 0$ of function $x(t)$ (Fig. 3.2). For the events that $0 < x < \infty$ and $a - \dot{x}dt \le x \le a$, where $\dot{x}$ is time derivative, the function $x(t)$ crosses level a from downwards at an instant τ, (for the case of upward sloping function). It is assumed that joint probability density function $p(x, \dot{x})$ is defined. The probability of both events is following

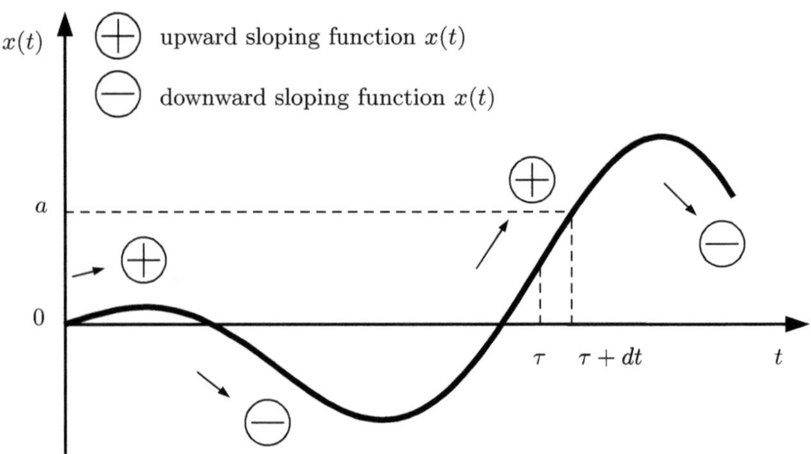

Fig. 3.2. A section of random history $x(t)$

$$\int\limits_{0}^{\infty} \int\limits_{a-\dot{x}dt}^{a} p(x,\dot{x})dx d\dot{x} \,. \tag{3.38}$$

Under the above assumptions a formula for mean number of level $x = a$ crossings N_a^+ in time unit is derived, first quoted by Rice in 1945 [80]

$$N_a^+ = \int\limits_{0}^{\infty} \dot{x}p(a,\dot{x})d\dot{x} \,. \tag{3.39}$$

Under the assumption that loading is a random history $x(t)$ with normal probability distribution of instantaneous values, the distribution of the mean number of a level crossing in second could be defined.

The function of probability density of a normal process takes the form

$$p(x) = \frac{1}{\sqrt{2\pi\mu_x}} \exp\left(-\frac{x^2}{2\mu_x}\right) \,. \tag{3.40}$$

It can be remarked that all derivatives of a normal random function have also normal probability distributions:

$$p(\dot{x}) = \frac{1}{\sqrt{2\pi\mu_{\dot{x}}}} \exp\left(-\frac{\dot{x}^2}{2\mu_{\dot{x}}}\right) \tag{3.41}$$

and

$$p(\ddot{x}) = \frac{1}{\sqrt{2\pi\mu_{\ddot{x}}}} \exp\left(-\frac{\ddot{x}^2}{2\mu_{\ddot{x}}}\right) \,. \tag{3.42}$$

Joint probability density function can be derived from the relationship

$$p(x, \dot{x}) = \frac{1}{2\pi\sqrt{\mu_x \mu_{\dot{x}}}} \exp\left[-\frac{1}{2}\left(\frac{x^2}{\mu_x} + \frac{\dot{x}^2}{\mu_{\dot{x}}} \right) \right]. \tag{3.43}$$

The variances of the random history $x(t)$ and its first $\dot{x}(t)$ and second $\ddot{x}(t)$ time derivatives are included in the formulae above. They are derived from power spectral density function $G_x(f)$:

$$\mu_x = \int_0^\infty G_x(f)df, \quad \mu_{\dot{x}} = \int_0^\infty G_x(f)f^2 df, \quad \mu_{\ddot{x}} = \int_0^\infty G_x(f)f^4 df. \tag{3.44}$$

Upon consideration of formula for k-th moment of power spectral density function

$$m_k = \int_0^\infty G_x(f)f^k df, \tag{3.45}$$

the expressions from (3.44) take the form:

$$\mu_x = m_0, \qquad \mu_{\dot{x}} = m_2, \qquad \mu_{\ddot{x}} = m_4. \tag{3.46}$$

By substitution of (3.43) into (3.39) the mean number of crossings in a time unit is derived for normal random history

$$N_a^+ = \frac{\sqrt{\mu_{\dot{x}}}}{\sqrt{\mu_x}} \exp\left(-\frac{a^2}{2\mu_x} \right). \tag{3.47}$$

On the basis of (3.47) the approximated mean number of oscillations N_0^+ (number of zero level crossing in secound) can be determined for narrow-band frequency history. To this end, the level must be assumed to take the value $a = 0$

$$N_0^+ = \sqrt{\frac{\mu_{\dot{x}}}{\mu_x}} = \sqrt{\frac{m_2}{m_0}}. \tag{3.48}$$

The distribution function of peaks (local extremes) is defined under formula

$$P(a) = P(x < a) = 1 - \frac{N_a^+}{N_0^+} \quad \text{for} \quad a > 0 \tag{3.49}$$

and, subsequently, probability density function

$$p(a) = \frac{dP(a)}{da} = -\frac{1}{N_0^+}\frac{dN_a^+}{da} \quad \text{for} \quad a > 0. \tag{3.50}$$

By differentiation of the relation (3.47) in the formula above, the probability density function of peaks is defined as

$$p(a) = \frac{a}{\mu_x} \exp\left(-\frac{a^2}{2\mu_x}\right). \tag{3.51}$$

The resulting distribution takes the form of the familiar Rayleigh distribution. It can be remarked that thus an analytical formula (3.51) is derived for the description of amplitude distribution for the considered case. There is no need to apply algorithmic methods of schematisation of random histories. The advantages of the procedure include simple and condensed formula shape, applicability in the analytical calculation for fatigue life determination and the reduction of computational time in comparison to various algorithms of the cycle counting methods. The drawback consists in the limited applicability due to the assumption that normal probability distribution of loading type with narrow-band frequency spectrum is under consideration.

3.5 Spectral Method of Fatigue Life Determination

The characteristic feature of the considered spectral method of fatigue life determination is application of the power spectral density function of stress or strain for estimation of amplitude probability distribution. It should be remembered that the PSD function does not sufficiently define stress or strain history and the information about the shape of probability distribution of the instantaneous values is indispensable. In the course of deriving life time formula, a hypothesis of the accumulation of fatigue damage is applied along with standard fatigue characteristics of materials, similar as in the cycle counting methods. The following subchapters present the assumptions and the process of spectral formulae derivation. The formulae apply two various fatigue characteristics, $(\sigma_a - N_f)$, which relates fatigue life to stress amplitude and $(\varepsilon_a - N_f)$, which relates fatigue life to strain amplitude.

3.5.1 Fatigue Life Calculation Based on Characteristics $(\sigma_a - N_f)$

For cyclic loading with constant amplitude, damage is calculated from the formula

$$D = \frac{n}{N_f}, \tag{3.52}$$

where: n – number of stress cycles with constant amplitude,

N_f – number of cycles to failure derived from Wöhler curve $(\sigma_a - N_f)$.

For variable amplitude loading the formula (3.52) takes the form

$$D = \sum_{i=1}^{k} D_i = \sum_{i=1}^{k} \frac{n_i}{N_{f\,i}}, \tag{3.53}$$

where: k – number of stress amplitude levels.

For the case of random histories it is difficult to determine explicitly the number of cycles with specific amplitude. For the purpose cycle counting algorithms, such as rain flow, range pairs, or histeresis loop are applied; however, they cannot be directly applied in the spectral methods. In order to overcome the problem it is assumed that random history is an ergodic stationary history with narrow-band frequency spectrum. For this case, local extrema (peaks) could be considered as amplitudes of history components[2]. If probability of the occurrence of peaks is denoted as $P(\sigma_a)$, n_i is the number of cycles with amplitude σ_{ai} during observation time T_o, the formula for n_i takes the form

$$n_i = n(\sigma_a) = M^+ T_o P(\sigma_a), \tag{3.54}$$

where: M^+ – expected number of peaks in a time unit.

By the application of formula (3.54) and replacing the summation from (3.53) with an integral in the range 0 to ∞, the following expression for damage in observation time T_o is derived

$$D(T_o) = \sum_{i=1}^{k} D_i = M^+ T_o \int_0^\infty \frac{p(\sigma_a)}{N_f(\sigma_a)} d\sigma_a , \tag{3.55}$$

where: $p(\sigma_a)$ – probability density function of peaks of stress history $\sigma(t)$,

$N_f(\sigma_a) = A\sigma_a^{-m}$ – number of cycles derived from the Wöhler curve.

By consideration of the fact that material fatigue life T expressed in seconds corresponds to the damage $D(T) = 1$ from the proportion:

$$\frac{1}{D(T_o)} = \frac{T}{T_o}, \tag{3.56}$$

the formula is derived

$$T = \frac{T_o}{D(T_o)} = \frac{1}{M^+ \int_0^\infty \frac{p(\sigma_a)}{N_f(\sigma_a)} d\sigma_a} . \tag{3.57}$$

If probability density function of peaks $p(\sigma_a)$ is described in terms of the Rayleigh distribution

$$p(\sigma_a) = \frac{\sigma_a}{\mu_\sigma} \exp\left(-\frac{\sigma_a^2}{2\mu_\sigma}\right), \tag{3.58}$$

the integral expression from (3.57) takes the form:

---

[2] Algorithmic ranges method described in ASTM standard [1] is equivalent with this approach.

$$\int_0^\infty \frac{p(\sigma_a)}{N_f(\sigma_a)} d\sigma_a = \int_0^\infty \frac{\frac{\sigma_a}{\mu_\sigma} \exp\left(\frac{-\sigma_a^2}{2\mu_\sigma}\right)}{A\sigma_a^{-m}} d\sigma_a = \frac{1}{A} \int_0^\infty \frac{\sigma_a}{\mu_\sigma} \exp\left(\frac{-\sigma_a^2}{2\mu_\sigma}\right) \sigma_a^m d\sigma_a \,,$$

$$(3.59)$$

where: $\mu_\sigma = m_0$ – variation of stress history $\sigma(t)$.

By analysis of the expression we can remark that the mathematical structure is similar to gamma function, also known as the Eulerian integral of second order [37]

$$\Gamma(z) = \int_0^\infty e^{-t} t^{z-1} dt \,.$$

$$(3.60)$$

The integral can be replaced with the gamma function after mathematical operations involving substitution of variables

$$t = \frac{\sigma_a^2}{2\mu_\sigma} \,, \qquad \sigma_a = (t2\mu_\sigma)^{\frac{1}{2}} \,, \qquad dt = \frac{\sigma_a}{\mu_\sigma} d\sigma_a \qquad (3.61)$$

hence,

$$\frac{1}{A} \int_0^\infty e^{-t} (t2\mu_\sigma)^{\frac{m}{2}} dt = \frac{(2\mu_\sigma)^{\frac{m}{2}}}{A} \int_0^\infty e^{-t} t^{\frac{m+2}{2}-1} dt \,.$$

$$(3.62)$$

Subsequently, the integral (3.59) is equal to

$$\int_0^\infty \frac{p(\sigma_a)}{N_f(\sigma_a)} d\sigma_a = \frac{(2\mu_\sigma)^{\frac{m}{2}}}{A} \Gamma\left(\frac{m+2}{2}\right) \,.$$

$$(3.63)$$

By referring back to fatigue life formula (3.57), the known Miles formula [59] is derived

$$T = \frac{1}{M^+ + \int_0^\infty \frac{p(\sigma_a)}{N_f(\sigma_a)} d\sigma_a} = \frac{A}{M^+ (2\mu_\sigma)^{\frac{m}{2}} \Gamma\left(\frac{m+2}{2}\right)} \,.$$

$$(3.64)$$

For stress history with narrow-band frequency spectrum the expected number of peaks per second M^+ is equal to the number of zero level crossings with positive slope N_0^+ or the dominant frequency f_0, i.e.

$$M^+ = N_0^+ = \sqrt{\frac{m_2}{m_0}} = f_0 \,.$$

$$(3.65)$$

The fact is often applied by the authors of papers [38, 103] for the determination of the expected number of cycles per time unit on the basis of the expected number zero level crossings, since the parameter is less sensitive to the noise occurring in the analysed signal.

During the derivation of fatigue life formula (3.64) it is assumed that amplitudes with ranges $(0, +\infty)$ participate in the process of damage accumulation. Damage accumulation is often applied only for amplitudes above a specified value in accordance with the conviction that amplitudes below some limit do not affect material damage considerably. The limit is associated with fatigue limit σ_{af}, and accumulation of damage is conducted in accordance with the modified Palmgren-Miner hypothesis. The hypothesis is illustrated in Fig. 3.3.

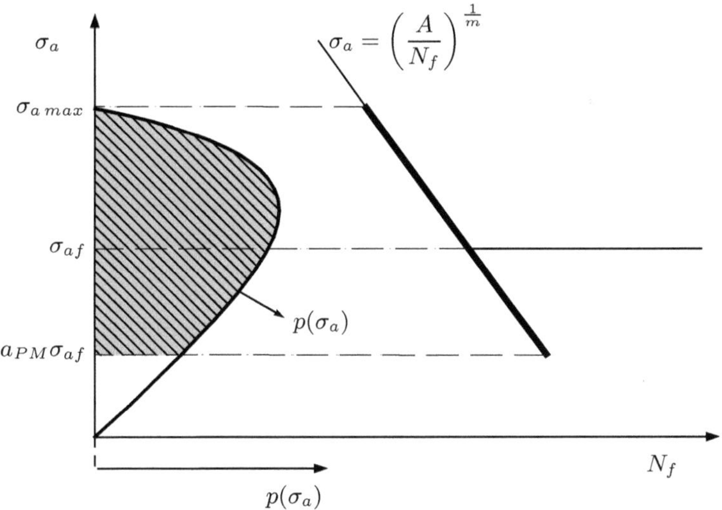

Fig. 3.3. Damage accumulation in accordance with modified Palmgren-Miner hypothesis. The bold line marks a section of fatigue characteristics participating in the process of damage accumulation

Damage for the postulated hypothesis can be derived from formula (2.66) in the cycle counting method. In the spectral method, the alteration of the lower limit of integration in formula (3.55) must be accounted for during derivation

$$D(T_o) = \sum_{i=1}^{k} D_i = M^{+}T_o \int\limits_{a_{PM}\sigma_{af}}^{\infty} \frac{p(\sigma_a)}{N_f(\sigma_a)} d\sigma_a , \qquad (3.66)$$

where: a_{PM} – coefficient accounting for amplitudes below fatigue limit σ_{af}.

By means of further transformations with substitutions from (3.61), the final life time formula is derived in the form

$$T = \frac{A}{M^+ (2\mu_\sigma)^{\frac{m}{2}} \; \Gamma\left(\dfrac{m+2}{2}, \dfrac{a_{PM}^2 \sigma_{af}^2}{2\mu_\sigma}\right)}, \tag{3.67}$$

where

$$\Gamma(z,a) = \int_a^\infty e^{-t} t^{z-1} dt \tag{3.68}$$

is the incomplete gamma function [37]. It can be remarked that the formula (3.67) is a restatement of Miles formula with coefficient $a_{PM} = 0$.

Haibach [29] postulates accounting for amplitudes below the fatigue limit σ_{af} by the modification of the Wöhler curve exponent during damage accumulation (Fig. 3.4)

$$D(T_o) = \sum_{i=1}^{k} \begin{cases} \dfrac{n_i}{N_G \left(\dfrac{\sigma_{af}}{\sigma_{ai}}\right)^m} & \text{for} \quad \sigma_{ai} \geq \sigma_{af} \\[4ex] \dfrac{n_i}{N_G \left(\dfrac{\sigma_{af}}{\sigma_{ai}}\right)^{m_H}} & \text{for} \quad \sigma_{ai} < \sigma_{af} \end{cases}, \tag{3.69}$$

where: $D(T_o)$ – damage determined for observation time T_o,
         n_i – number of cycles with amplitude σ_{ai},
         σ_{af} – fatigue limit,
         $m_H = 2m - 1$ – modifies sloping of fatigue characteristics for amplitudes below fatigue limit.

By application of the probability density distribution function of amplitude $p(\sigma_a)$, the conditional sum from equation (3.69) could be replaced with the sum of two integrals

$$D(T_o) = M^+ T_o \int_{\sigma_{af}}^\infty \frac{p(\sigma_a)}{N_f(\sigma_a)} d\sigma_a + M^+ T_o \int_0^{\sigma_{af}} \frac{p(\sigma_a)}{N_{f\,H}(\sigma_a)} d\sigma_a, \tag{3.70}$$

where: $N_{f\,H}(\sigma_a) = A\sigma_a^{-m_H}$ – number of cycles form Haibach chart for amplitudes below fatigue limit (Fig. 3.4).

Under the assumption that amplitudes are defined by means of the Rayleigh distribution, the integrals from (3.70) take the simplified form:

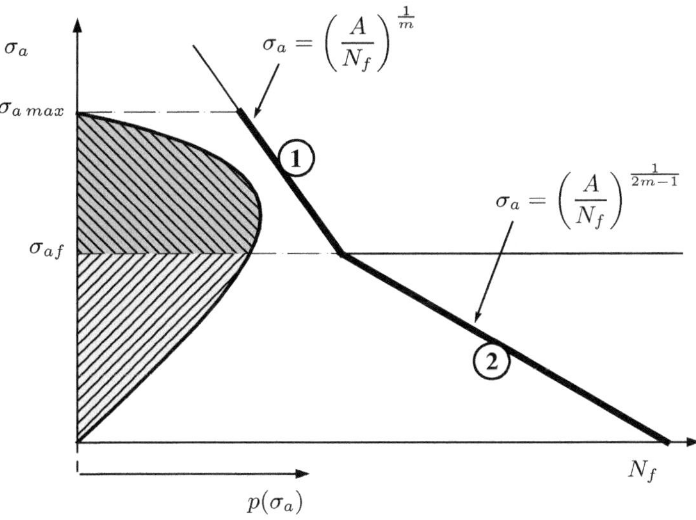

Fig. 3.4. Fatigue accumulation in accordance with Haibach hypothesis [29]. Bold line 1 marks a section of Wöhler curve taking part in damage accumulation for amplitudes $\sigma_a \geq \sigma_{af}$. Line 2 marks modified characteristics with slope coefficient $m_H = 2m - 1$ for amplitudes $\sigma_a < \sigma_{af}$

$$\int_{\sigma_{af}}^{\infty} \frac{p(\sigma_a)}{N_f(\sigma_a)}\,d\sigma_a = \frac{(2\mu_\sigma)^{\frac{m}{2}}}{A}\,\Gamma\left(\frac{m+2}{2}, \frac{\sigma_{af}^2}{2\mu_\sigma}\right), \tag{3.71}$$

$$\int_{0}^{\sigma_{af}} \frac{p(\sigma_a)}{N_{f\,H}(\sigma_a)}\,d\sigma_a = \frac{(2\mu_\sigma)^{\frac{m_H}{2}}}{A_H}\left[\Gamma\left(\frac{m_H+2}{2}\right) - \Gamma\left(\frac{m_H+2}{2}, \frac{\sigma_{af}^2}{2\mu_\sigma}\right)\right], \tag{3.72}$$

where: $A_H = N_G\sigma_{af}^{m_H}$ – constant derived from the Haibach chart.

By the application of the Haibach linear hypothesis of damage accumulation the fatigue life formula is derived by the transformation of expression (3.70) under the assumptions that $D(T_o) = 1$ and $T_o = T$

$$T = \frac{A}{M^+ (2\mu_\sigma)^{\frac{m}{2}}\,\Gamma\left(\dfrac{m+2}{2}, \dfrac{\sigma_{af}^2}{2\mu_\sigma}\right)} + \frac{A_H}{M^+ (2\mu_\sigma)^{\frac{m_H}{2}}\left[\Gamma\left(\dfrac{m_H+2}{2}\right) - \Gamma\left(\dfrac{m_H+2}{2}, \dfrac{\sigma_{af}^2}{2\mu_\sigma}\right)\right]}. \tag{3.73}$$

Corten and Dolan [20, 95] postulate the application of secondary fatigue charts with a different exponent $q_{CD} \neq m$ for fatigue computation. The hypothesis is illustrated in Fig. 3.5.

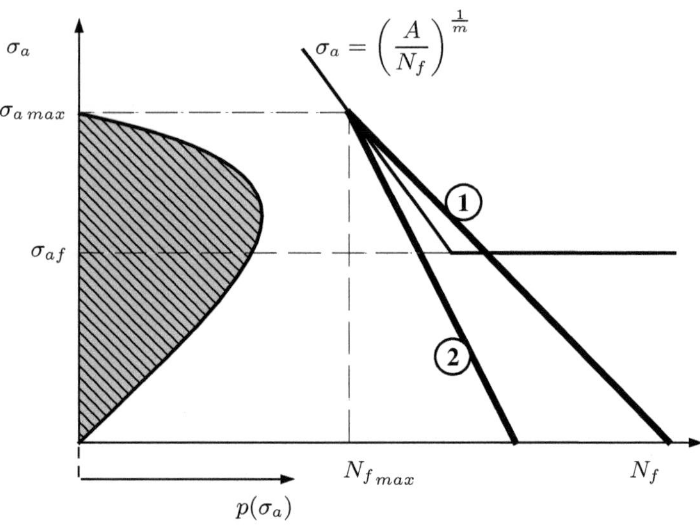

Fig. 3.5. Damage accumulation in accordance with Corten-Dolan hypothesis. Bold lines mark fatigue characteristics participating in damage accumulation process. Various slope of fatigue characteristics depending on exponent q_{CD} ($q_{CD} < m$ for line 1, $q_{CD} > m$ for line 2) are gained

The secondary fatigue chart has a point of contact with Wöhler chart for stress amplitude $\sigma_a = \sigma_{amax}$. Damage accumulation is performed for the entire amplitude spectrum ($0 < \sigma_a \leq \sigma_{a\,max}$). Damage for the hypothesis is expressed with the formula

$$D(T_o) = \sum_{i=1}^{k} \frac{n_i}{N_{f\,max}} \left(\frac{\sigma_{ai}}{\sigma_{a\,max}}\right)^{q_{CD}}, \tag{3.74}$$

where: $\sigma_{a\,max}$, $N_{f\,max}$ – maximum amplitude of a cycle in realisation T_o and the corresponding number of cycles to failure derived from the Wöhler curve.

In order to derive spectral damage formula under the assumption that random stress history $\sigma(t)$ has normal distribution with narrow-band frequency spectrum and by application of Corten-Dolan hypothesis, the sum in equation (3.74) must be replaced with an integral

$$D(T_o) = M^+ T_o \int\limits_0^\infty \frac{p(\sigma_a)}{N_{f\,max}} \left(\frac{\sigma_a}{\sigma_{a\,max}} \right)^{q_{CD}} d\sigma_a \,, \tag{3.75}$$

and the Rayleigh distribution may be applied (3.58) for the description of amplitude probability density function $p(\sigma_a)$

$$D(T_o) = M^+ T_o \int\limits_0^\infty \frac{\dfrac{\sigma_a}{\mu_\sigma} \exp\left(-\dfrac{\sigma_a^2}{2\mu_\sigma} \right)}{A\sigma_{a\,max}^{-m}} \left(\frac{\sigma_a}{\sigma_{a\,max}} \right)^{q_{CD}} d\sigma_a \,. \tag{3.76}$$

After mathematical transformations performed similarly as for the derivation of (3.64), the damage formula takes the form

$$D(T_o) = \frac{M^+ T_o \,(2\mu_\sigma)^{\frac{q_{CD}}{2}}}{A\sigma_{a\,max}^{q_{CD}-m}} \Gamma\left(\frac{q_{CD} + 2}{2} \right) \,. \tag{3.77}$$

Under an assumption of damage $D(T_o) = 1$ for observation time equal to fatigue life $T_o = T$, the following formula is obtained

$$T = \frac{A\sigma_{a\,max}^{q_{CD}-m}}{M^+ (2\mu_\sigma)^{\frac{q_{CD}}{2}} \Gamma\left(\dfrac{q_{CD} + 2}{2} \right)} \,. \tag{3.78}$$

Serensen and Kogayev [51, 88, 95] postulate damage accumulation on the basis of secondary fatigue chart. It is derived by modification of the Wöhler curve by accounting for coefficient b_{SK} (Fig. 3.6). After the adoption of linear Serensen-Kogayev hypothesis, the damage formula takes the form:

$$D(T_o) = \frac{1}{b_{SK}A} \sum_{i=1}^{k} n_i \sigma_{ai}^m \qquad \text{for} \quad \sigma_{ai} \geq a_{SK}\sigma_{af} \,, \tag{3.79}$$

where: $b_{SK} = \dfrac{\sum\limits_{i=1}^{k} \sigma_{ai} t_i - a_{SK}\sigma_{af}}{\sigma_{a\,max} - a_{SK}\sigma_{af}}$ – Serensen-Kogayev coefficient characterizes amplitude spectrum of random loading (applicable for $b_{SK} > 0.1$, $\sigma_{a\,max}/\sigma_{af} > 1$ and $\dfrac{1}{\sigma_{a\,max}} \sum\limits_{i=1}^{k} \sigma_{ai} t_i > 0.5$),

$a_{SK} \in \langle 0, \ldots, 1 \rangle$ – coefficient accounting for amplitudes below fatigue limited, for presentations $a_{SK} = 0.5$ was adopted,

$\sigma_{a\,max}$ – maximum amplitude of counted cycles,

$t_i = \dfrac{n_i}{\sum\limits_{i=1}^{k} n_i}$ – frequency of levels σ_{ai} occurrence for observation time T_o.

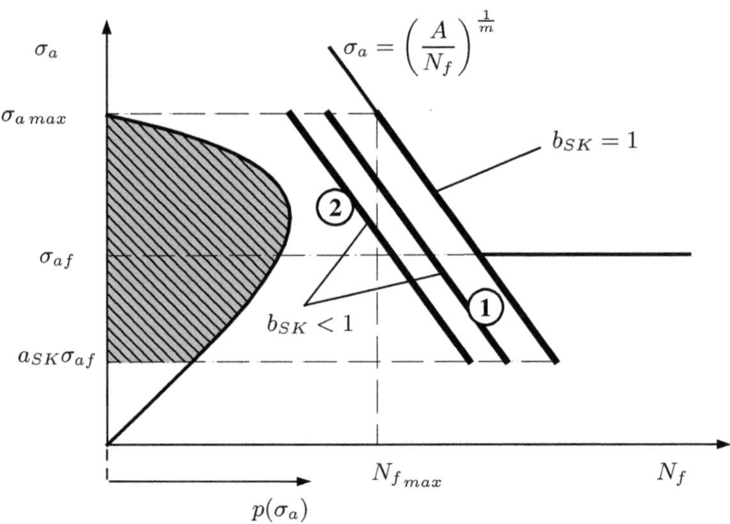

Fig. 3.6. Serensen-Kogayev hypothesis of damage accumulation. The bold lines represent fatigue characteristics participating in damage accumulation process (b_{SK} coefficient is higher for line 1 than for line 2)

By comparison of formulae (3.66) and (3.79) for damage determination $D(T_o)$ a remark that the difference consists in coefficient b_{SK} could be stated. By application of the premises for the statement of damage formula (3.66), the expression accounting for amplitudes below fatigue limit takes the form

$$D(T_o) = \frac{M^+ T_o}{b_{SK}} \int\limits_{a_{SK}\sigma_{af}}^{\infty} \frac{p(\sigma_a)}{N_f(\sigma_a)} d\sigma_a \,, \tag{3.80}$$

$$b_{SK} = \frac{\int\limits_{a_{SK}\sigma_{af}}^{\infty} \sigma_a p(\sigma_a) d\sigma_a - a_{SK}\sigma_{af}}{\sigma_{a\,max} - a_{SK}\sigma_{af}} \,. \tag{3.81}$$

The formula for fatigue life T and coefficient b_{SK} could be restated in the light of the spectral aspect under an assumption that amplitudes are approximated by means of the Rayleigh distribution (3.51) [49]

$$T = \frac{A}{M^+ (2\mu_\sigma)^{\frac{m}{2}} \Gamma\left(\dfrac{m+2}{2}, \dfrac{a_{SK}^2 \sigma_{af}^2}{2\mu_\sigma}\right)} \,, \tag{3.82}$$

where

$$b_{SK} = \frac{\sqrt{2\mu_\sigma}\,\Gamma\left(\dfrac{3}{2},\dfrac{a_{SK}^2\sigma_{af}^2}{2\mu_\sigma}\right)\exp\left(\dfrac{a_{SK}^2\sigma_{af}^2}{2\mu_\sigma}\right) - a_{SK}\sigma_{af}}{\sigma_{a\,max} - a_{SK}\sigma_{af}}\,. \tag{3.83}$$

3.5.2 Fatigue Life Calculation Based on Characteristics $(\varepsilon_a - N_f)$

The following formula is derived under the same assumptions as in the preceding subchapter, while fatigue characteristics $(\sigma_a - N_f)$ is now replaced with the characteristics $(\varepsilon_a - N_f)$, as described by the Manson-Coffin-Basquin equation, which relates the total strain amplitude to the number of cycles to failure

$$\varepsilon_a = \varepsilon_{ae} + \varepsilon_{ap} = \frac{\sigma'_f}{E}(2N_f)^b + \varepsilon'_f(2N_f)^c\,. \tag{3.84}$$

Formula (3.57) was applied for the expression of damage as the functions relating to strain amplitude ε_a

$$T = \frac{1}{M + \displaystyle\int\limits_0^\infty \frac{p(\varepsilon_a)}{N_f(\varepsilon_a)}d\varepsilon_a}\,, \tag{3.85}$$

where: $p(\varepsilon_a)$ – probability density function of peaks of strain history $\varepsilon(t)$ with narrow-band frequency spectrum,

$N_f(\varepsilon_a)$ – function of cycle number on the basis of the Manson-Coffin-Basquin equation.

The function $N_f(\varepsilon_a)$ could not be directly determined from the formula (3.84). Therefore, fatigue life T must be computed from formula (3.85) by means of a numerical method. In a number of cases the process of material damage is not associated with plastic strain with reference to the large number of cycles to failure [66]. In that case, plastic strain could be neglected in (3.84) and the formula takes the form

$$\varepsilon_a = \varepsilon_{ae} = \frac{\sigma'_f}{E}(2N_f)^b\,, \tag{3.86}$$

after the restatement of which the analytical expression of cycle number $N_f(\varepsilon_a)$ takes the form

$$N_f(\varepsilon_a) = \frac{1}{2}\left(\frac{\varepsilon_a E}{\sigma'_f}\right)^{\frac{1}{b}}\,. \tag{3.87}$$

By application of the function and deriving the formula further, similar as in subchapter (3.5.1), the formula takes the form

$$T = \frac{(2\mu_\varepsilon)^{\frac{1}{2b}}}{M + 2\left(\dfrac{\sigma'_f}{E}\right)^{\frac{1}{b}}\Gamma\left(1 - \dfrac{1}{2b}\right)}\,, \tag{3.88}$$

where: μ_ε – variance of strain history $\varepsilon(t)$.

Unfortunately the neglect of the plastic strain imposes limitations on the applicability of the formula for determination of life time with reference to low-cycle fatigue, in case of which plastic strain is larger in comparison to elastic strain. In order to overcome the problem, it was postulated that fatigue characteristics $(\varepsilon_a - N_f)$ as defined in formula (3.84) could be replaced with a family of tangent sections to Manson-Coffin-Basquin chart in a double logarithmic system. Fig. 3.7 presents three sample tangents and marking the points of intersection. Under the assumptions the fatigue life formula takes the form

$$T = \cfrac{1}{M + \sum\limits_{k=1}^{ns}\left[\int\limits_{\varepsilon_{lk}}^{\varepsilon_{uk}} \dfrac{p(\varepsilon_a)}{N_{f_k}(\varepsilon_a)}d\varepsilon_a\right]} , \qquad (3.89)$$

where: ns – number of tangent sections,

ε_{lk}, ε_{uk} – low and high limit of integration for k-th tangent,

$N_{f_k}(\varepsilon_a)$ – function defining the number of cycles for k-th tangent section from characteristics $(\varepsilon_a - N_f)$.

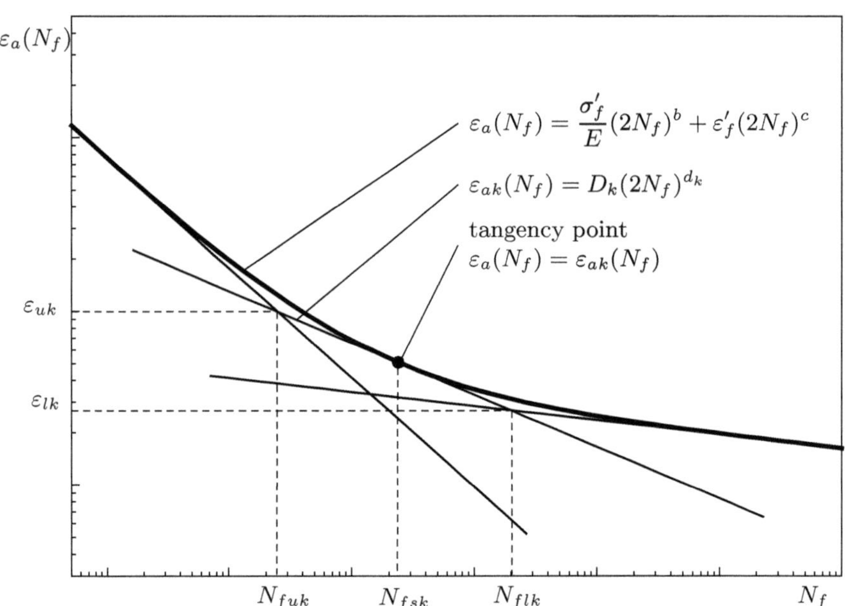

Fig. 3.7. Manson-Coffin-Basquin curve approximated with three tangent sections in a double algorithmic system (details in text)

Under the assumption that the k-th tangent to characteristics $(\varepsilon_a - N_f)$ is defined by the function

$$\varepsilon_{ak}(N_f) = D_k(2N_f)^{d_k}\,, \qquad (k = 1, \ldots, ns)\,, \tag{3.90}$$

where: D_k, d_k – parameters,

the inverse function could be defined. It is applied for determination of the number of cycles to failure in the range $\langle \varepsilon_{lk}, \varepsilon_{uk} \rangle$

$$N_{fk}(\varepsilon_a) = \frac{1}{2} \left(\frac{\varepsilon_a}{D_k} \right)^{\frac{1}{d_k}}\,, \qquad \varepsilon_a \in \langle \varepsilon_{lk}, \varepsilon_{uk} \rangle \,. \tag{3.91}$$

By consideration of the integral from formula (3.89) in the range ε_{lk}, ε_{uk} and under the assumption that amplitude probability density function $p(\varepsilon_a)$ is described by means of the Rayleigh distribution (3.51) (random history with normal distribution strain and with narrow-band frequency spectrum) the formula assumes the form

$$\int_{\varepsilon_{lk}}^{\varepsilon_{uk}} \frac{p(\varepsilon_a)}{N_{fk}(\varepsilon_a)} d\varepsilon_a = \int_{\varepsilon_{lk}}^{\varepsilon_{uk}} \frac{\varepsilon_a}{\mu_\varepsilon} e^{\frac{-\varepsilon_a^2}{2\mu_\varepsilon}} 2 \left(\frac{D_k}{\varepsilon_a} \right)^{\frac{1}{d_k}} d\varepsilon_a\,. \tag{3.92}$$

In order to perform further transformations, the integrand expression may be replaced by the following substitutions:

$$t = \frac{\varepsilon_a^2}{2\mu_\varepsilon}\,, \qquad \varepsilon_a = \sqrt{t2\mu_\varepsilon}\,, \qquad dt = \frac{\varepsilon_a}{\mu_\varepsilon} d\varepsilon_a\,,$$
$$t_{lk} = \frac{\varepsilon_{lk}^2}{2\mu_\varepsilon}\,, \qquad t_{uk} = \frac{\varepsilon_{uk}^2}{2\mu_\varepsilon}\,. \tag{3.93}$$

After the substitution

$$\int_{\varepsilon_{lk}}^{\varepsilon_{uk}} \frac{p(\varepsilon_a)}{N_{fk}(\varepsilon_a)} d\varepsilon_a = 2 \left(\frac{D_k}{\sqrt{2\mu_\varepsilon}} \right)^{\frac{1}{d_k}} \int_{t_{lk}}^{t_{uk}} e^{-t} t^{-\frac{1}{2d_k}} dt\,. \tag{3.94}$$

It could be remarked that the mathematical structure of the expression is similar to the incomplete gamma function [37]

$$\Gamma(z, a) = \int_{a}^{\infty} e^{-t} t^{z-1} dt\,. \tag{3.95}$$

Aware of the fact that

$$\int_{x_1}^{x_2} f(x)dx = \int_{x_1}^{\infty} f(x)dx - \int_{x_2}^{\infty} f(x)dx\,, \qquad (x_1 < x_2)\,, \tag{3.96}$$

the integral in equation (3.94) is substituted with the difference of two incomplete gamma functions and by substitution into (3.85), the formula is restated in the following form

$$T = \cfrac{1}{2M + \sum\limits_{k=1}^{ns} \left(\cfrac{D_k}{\sqrt{2\mu_\varepsilon}} \right)^{\frac{1}{d_k}} \left[\Gamma\left(1 - \cfrac{1}{2d_k}, t_{lk}\right) - \Gamma\left(1 - \cfrac{1}{2d_k}, t_{uk}\right) \right]} . \qquad (3.97)$$

The simple and concise form, the possibility of easy conversion into a program and the applicability in the determination of fatigue life with regard to low and high cycle numbers constitute the combined advantages of the formula (3.97). The disadvantage is the limited applicability due to the assumptions concerning the character of random strain history. Formula (3.97) could be generalised into the case of loading with arbitrary spectrum frequency band by the application of correction coefficient λ (2.30), similar as in Wirsching [103] for each tangent section.

During the application of the postulated formula the problems of adequate selection of tangent section number $N_{fk}(\varepsilon_a)$ and determination of the parameters D_k and d_k are encountered. From the simulations it results that in a majority of instances the number of tangent sections need not exceed 50 in order to ensure that life time by the method by formula (3.97) and by the numerical method (3.85) are similar. Figures 3.8 to 3.11 present error resulting from the approximation of the Manson-Coffin-Basquin chart with various numbers of tangent sections ($k = 2, 3, 9, 50$). The error is determined from formula

$$N_{f\,error} = \frac{N_{f\,num} - N_{f\,k}}{N_{f\,num}} 100\% , \qquad (3.98)$$

which defines per cent of difference between the number of cycles determined by means of the numerical method (estimation accuracy $N_{f\,num}(\varepsilon_a) = N_f(\varepsilon_a) \pm 2.2204 \cdot 10^{-16}$ performed by means of `fzero` function from `MATLAB` programme [58]) and the analytical application of k tangent sections characteristics. It could be remarked that the estimation error is dependent on the number of tangent sections, and does not exceed 1% for $k = 50$ (Fig. 3.11).

In order to present the effect of estimation error of the cycle number on fatigue life, simulations are performed for the uniform and Rayleigh amplitude distributions in relation to the number of assumed tangent sections $k = 2, \ldots, 100$ (Fig. 3.12). The calculations apply the Palmgren-Miner hypothesis of damage accumulation. It is remarked that for an arbitrary number of tangent sections the resulting damage is higher than for the case of direct determination by the Manson-Coffin-Basquin characteristics. This produces safe underestimation of fatigue life. In this case the conclusion that tangent section number $k = 50$ enables sufficient accuracy of calculation.

The parameters of k-th tangent section are derived from the adopted number of cycles N_{fsk}, which includes a point of tangency with the Manson-Coffin-Basquinne chart $(\varepsilon_a - N_f)$. A solution to a system of equations provided with

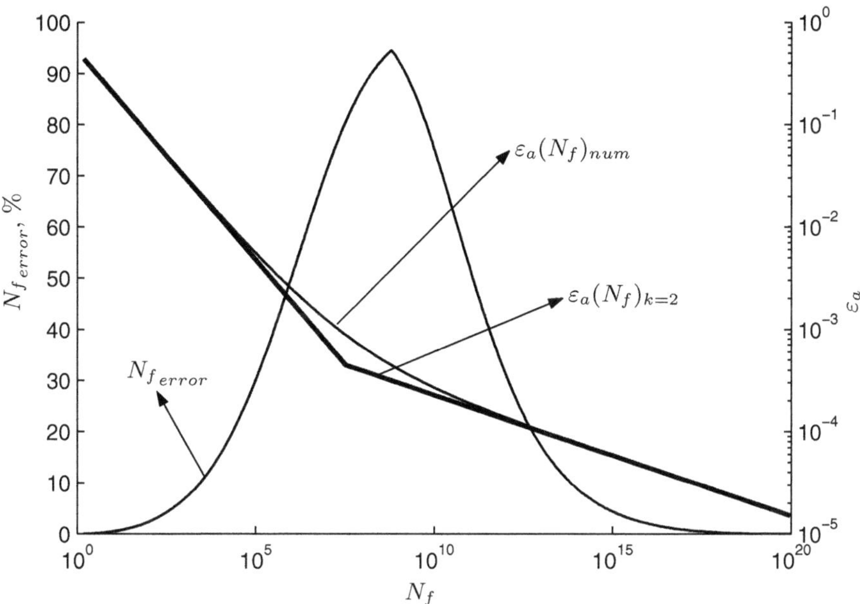

Fig. 3.8. Error of cycle number N_f determination for approximation of Manson-Coffin-Basquin fatigue characteristics with $k = 2$ tangent sections

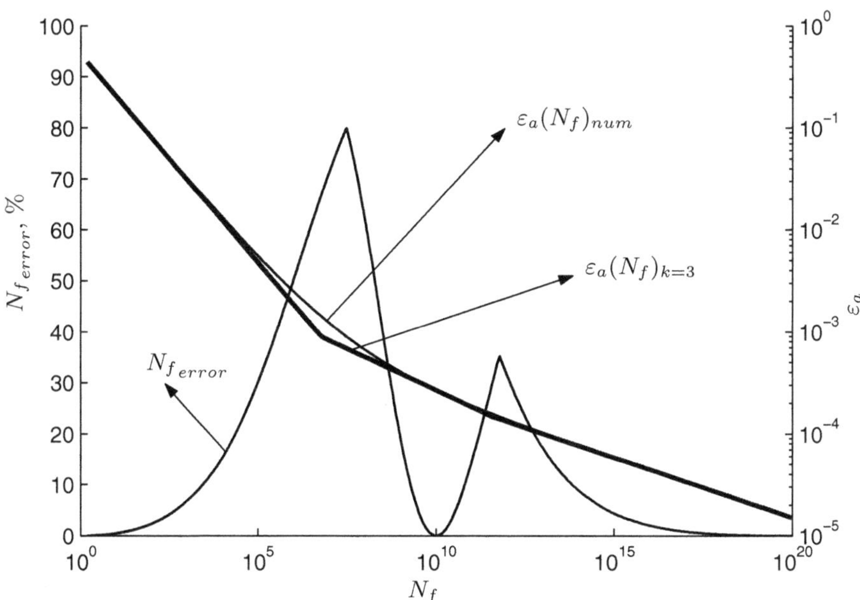

Fig. 3.9. Error of cycle number N_f determination for approximation of Manson-Coffin-Basquin fatigue characteristics with $k = 3$ tangent sections

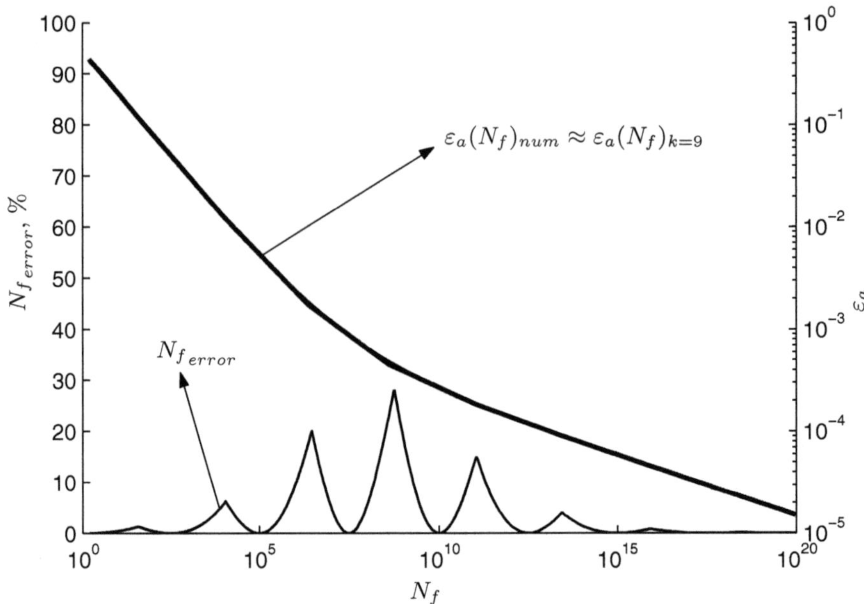

Fig. 3.10. Error of cycle number N_f determination for approximation of Manson-Coffin-Basquin fatigue characteristics with $k = 9$ tangent sections

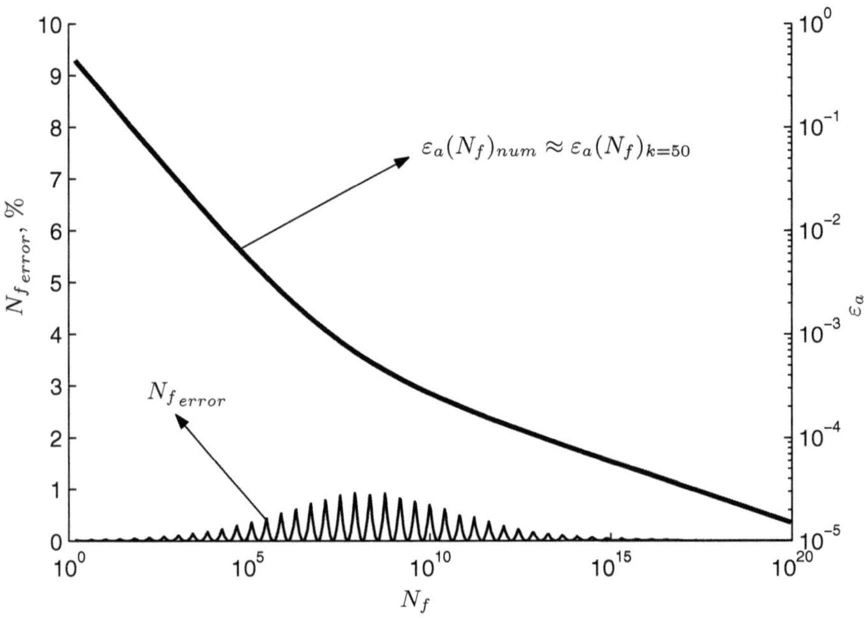

Fig. 3.11. Error of cycle number N_f determination for approximation of Manson-Coffin-Basquin fatigue characteristics with $k = 50$ tangent sections

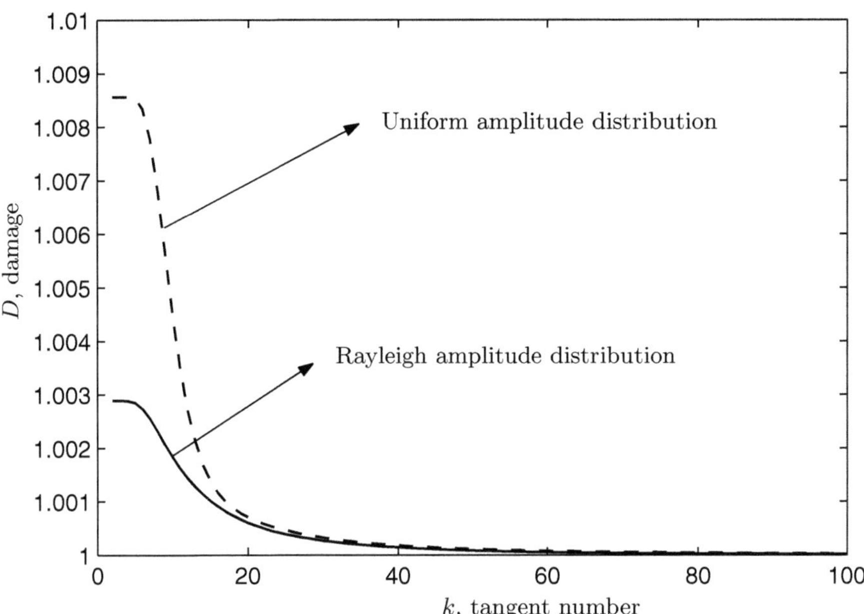

Fig. 3.12. Damage in the function of tangent section number. Broken line marks uniform amplitude distribution and full line – Rayleigh amplitude distribution

suitable relations

$$\begin{cases} \left.\dfrac{\partial \varepsilon_a(N_f)}{\partial N_f}\right|_{N_{fsk}} = \left.\dfrac{\partial \varepsilon_{ak}(N_f)}{\partial N_f}\right|_{N_{fsk}} , \\[2mm] \varepsilon_a(N_{fsk}) = \varepsilon_{ak}(N_{fsk}) \end{cases} \tag{3.99}$$

which served for the derivation of damage

$$D_k = \frac{\sigma'_f}{E}\left(2N_{fsk}\right)^{\frac{\varepsilon_{ap}(b-c)}{\varepsilon_a}} + \varepsilon'_f\left(2N_{fsk}\right)^{\frac{\varepsilon_{ae}(c-b)}{\varepsilon_a}} , \tag{3.100}$$

$$d_k = \frac{\varepsilon_{ae}\,b + \varepsilon_{ap}\,c}{\varepsilon_a} , \tag{3.101}$$

where: N_{fsk} – cycle for which equation $N_{fk}(\varepsilon_a) = N_f(\varepsilon_a)$ is fulfilled,
     $\varepsilon_a, \varepsilon_{ae}, \varepsilon_{ap}$ – amplitude of total, elastic, and plastic strain, respectively derived from (3.84) for $N_f = N_{fsk}$.

Strain amplitude ε_{uk} and ε_{lk} is determined from formulae

$$\varepsilon_{lk} = D_k\left(\frac{D_k}{D_{k+1}}\right)^{\frac{d_k}{d_{k+1}-d_k}} , \qquad \varepsilon_{uk} = D_k\left(\frac{D_k}{D_{k-1}}\right)^{\frac{d_k}{d_{k-1}-d_k}} . \tag{3.102}$$

The application of the method for the determination of fatigue life must involve the following stages:

- determination of tangent section number (recommended $ns = 50$),
- determination of point of tangency for every tangent section (total cycle number N_{fsk}) (logarithmic division in the range 0.5 to 10^{12} cycles is recommended),
- derivation of strain amplitudes ε_a, ε_{ae}, ε_{ap} for cycle number N_{fsk} on the basis of (3.84),
- derivation of coefficients D_k (3.100) and d_k (3.101) for each tangent section,
- determination of points of intersection ε_{lk} and ε_{uk} for adjoin tangent section (for extreme tangent sections $\varepsilon_{u\,1} = D_1$ and $\varepsilon_{l\,ns} = 0$),
- calculation of variance of strain history $\mu_\varepsilon = m_0$,
- calculation of integration limits t_{lk} and t_{uk} (3.93),
- calculation of fatigue life T from formula (3.97).

4

Algorithm of Spectral Method for Evaluation of Fatigue Life

For the case of multiaxial random loading the calculation of fatigue life consists in the reduction of triaxial stress or strain states into an equivalent uniaxial one by the application of suitable criteria of fatigue failure [53, 54, 75]. In the spectral methods, a matrix of power spectral density functions (3.14) defines stress or strain states. The reduction of multiaxial loading state may be conducted directly on power spectral density functions. The power spectral densities of equivalent stress or strain gained as a result of the application of multiaxial fatigue failure criteria are subsequently employed in the same way as the functions defined on the basis of uniaxial tension-compression testing [28]. This approach implements the familiar and experimentally verified methods of determination of fatigue life associated with simple loading states.

On the basis of the review of state-of-the-art in research and experience an algorithm for the determination of fatigue life under multiaxial loading is currently postulated with the application of the spectral method (Fig. 4.1). The following subchapters include the description of the particular blocks comprised in the algorithm.

4.1 Block 1 – Loading

The input data for the calculation of fatigue life are matrices of power spectral density functions of random stress or strain tensor (3.14). The functions could be derived by:

1. Measurement of strain in structural components under service conditions by means of spectrum analyser.
2. Calculation of real or designed structures by means of the Finite Element Method (FEM), Boundary Element Method (BEM) or Finite Difference Method (FDM) by application of analysis of frequency response of a system [30, 62, 74].

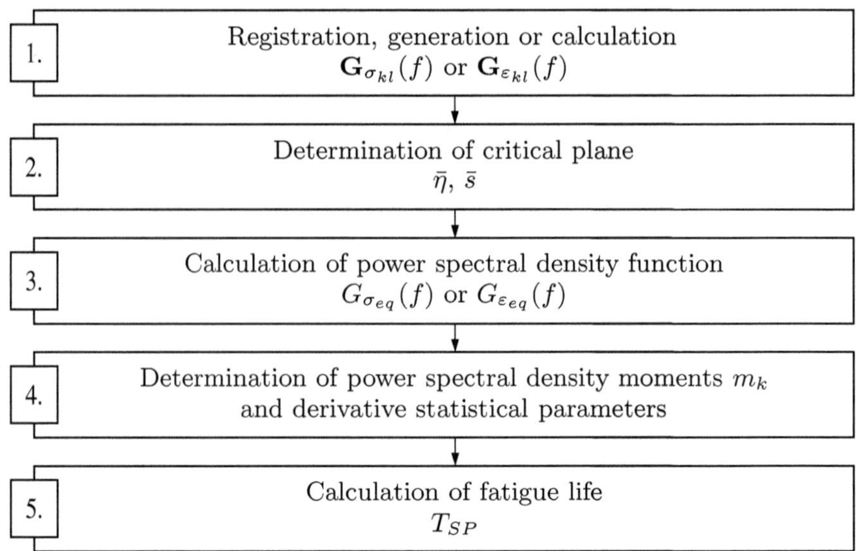

Fig. 4.1. Algorithm of fatigue life determination by spectral method for the multiaxial random loading defined with a matrix of power spectral density functions of stress or strain tensor components

3. Calculation of the PSD function of generated random sequences[1] with adaptable probabilistic characteristics corresponding to service conditions or anticipated circumstances.
4. Selection of parameters for the definition of the shape of the empirical PSD functions derived on the basis of a statistical elaboration of the identified physical phenomena[2].

The input quantity defining the loading state also includes the assumed probability distribution of instantaneous values of stress or strain tensor components.

4.2 Block 2 – Determination of Critical Plane Position

The determination of the expected position of critical fatigue fracture plane constitutes an important part of the algorithm of life time calculation. One

---

[1] For the purposes of fatigue life the estimators of auto- and cross-spectral density functions are most commonly determined by application of the fast Fourier transform with the Welch method [89, 101].

[2] Bibliography refers to only two instances in which the functions are applicable: longitudinal component of wind velocity [32] and sea surface rise [14, 30]. A considerable limitation to the method is the missing statement of the technique of generation of cross-PSD functions.

of the fundamental factors applied for the determination of the critical plane is the stress or strain state of the material. The position is most commonly determined by the definition of direction cosines l_n, m_n and n_n $(n = \eta, s)$ of unit vectors $\bar{\eta}$ and $\bar{s}$ occurring in the criteria of multiaxial fatigue failure, where $\bar{\eta}$ is orthogonal and $\bar{s}$ is tangent to the critical plane. Application of the following two methods is postulated for the determination of the expected position of the critical plane.

1. *Damage accumulation method*[3], discussed in [50, 51] consists in the determination of fatigue damage for the possible positions of the critical plane and the selection of the position for which the damage is at the maximum. As a result, fatigue life is determined besides the position of the critical plane.
2. *Variance method.* It is currently popular and offers good results [8, 9, 48]. Its application is not recommended for the instances when the components of stress or strain tensor have different probability distributions of instantaneous values [92]. In the variance method it is assumed that the planes for which the equivalent stress or strain variance reaches maximum in accordance with the selected criterion of multiaxial fatigue failure, are critical to the material and one of them might involve fatigue failure. The search for variance maximum $\mu_{x_{eq}}$ for the analysed stress or strain state consists in the seeking the maximum value of the expression

$$\mu_{x_{eq}} = \sum_{k=1}^{6}\sum_{l=1}^{6} a_k a_l \mu_{x_{kl}}, \tag{4.1}$$

where: $a_k, a_l = f(l_n, m_n, n_n, c_p)$ – non-linear functions of directional
$\qquad\qquad\qquad\qquad\qquad\qquad$ cosines (Table 3.1),
$\quad c_p \qquad\qquad\qquad\qquad$ – material constants.

The elements $\mu_{x_{kl}}$ of covariance matrix is calculated directly from random histories of component stress or strain state, although they could also be determined from the matrix of power spectral density (3.14)

$$\boldsymbol{\mu_x} = \mathrm{Re}\left[\int_0^{\infty} \mathbf{G_x}(f)df\right]. \tag{4.2}$$

In the case of the above methods, the successful application depends on the selection of an appropriate criterion of multiaxial fatigue failure and discretisation step of angles change unit vectors $\bar{\eta}$ and $\bar{s}$. In this book the method of maximum variance was employed due to its versatile use both in the spectral

---

[3] In the spectral method damage accumulation is not realised directly as iterative process familar from the cycle counting method. Damage is calculated from integral (3.57).

and in the cycle counting methods for fatigue life determination. A comparison of the two methods follows in the current book. Besides, the variance method is easily transformable into a computer programme and the algorithm is marked with a shorter calculation time in comparison to the method of damage accumulation.

4.3 Block 3 – Calculation of Power Spectral Density Function of Equivalent Stress or Strain

Block 3 is devoted to determination of power spectral density function of the equivalent stress or strain from formula (3.35). The formula could only be applied on condition that the equivalent value is defined as a linear combination of suitable components of stress or strain tensor according to the criteria of multiaxial fatigue failure.

4.4 Block 4 – Statistical Parameters

The statistical parameters defining the history of equivalent stress or strain are directly determined from moments of power spectral density of the equivalent quantities

$$m_k = \int_0^\infty f^k G_{eq}(f) df \, . \tag{4.3}$$

4.5 Block 5 – Calculation of Fatigue Life

The calculated fatigue life (3.57) is the time, commonly expressed in seconds, of material or structural component failure with 50% probability. The spectral method applies various formulae adopted for the particular characteristics of fatigue loading [11]. The following groups are distinguished in accordance with a general classification:

1. Spectral formulae for loading history with normal probability distribution,
   1.a. narrow-band frequency [19, 42, 59, 69],
   1.b. narrow-band frequency spectrum with an additional harmonic component or impulse loading [26, 33, 84],
   1.c. broad-band frequency spectrum [15, 17, 21, 38, 79, 100, 103],
2. Spectral formulae for loading history with probability distribution other than normal,
   2.a. narrow-band frequency spectrum [85],
   2.b. broad-band frequency spectrum [39, 43, 86].

It is remarked that the suitable selection of spectral formula depends on the loading character. In practice, the formulae are most commonly derived for the particular loading types or structures (drilling rigs, wind power plants, body plating, etc.). For the instances when the loading type (especially frequency width) is not determined, the application of broad frequency spectrum formulae is recommended (1.c. or 2.b.) due to the larger versatility. The application of spectral formulae groups 1.a. and 1.c. in this book results from the normal probability distribution of loading analysed in the simulation and experimental tests presented in the following chapters.

5

Simulations

5.1 Algorithm, Assumptions and Objectives of Simulations

A number of simulations of multiaxial random loading have been performed for improved comprehension and analysis of spectral method. They serve for comparison of the applied and known from literature cycle counting method with the postulated generalisation of the spectral method. Upon the extensive study of the algorithms of the two methods it is concluded that some calculation blocks include common features, which are subsequently established. Figure 5.1 presents block diagrams of two methods and focuses on the common features of the methods. The results of calculations of the two algorithms are compared in the distinguished blocks PWO1, PWO2 and PWO3.

The assumptions of the conditions for the simulations are established:

1. Generated stress or strain tensors have normal probability distribution for narrow- and broad-band frequency spectra;
2. Suitable stress or strain tensor components have different coefficients of cross-correlation ($r = 1$, $r = -1$ and $r \approx 0$);
3. Analysed stress states:
    - biaxial tension-compression,
    - uniaxial tension-compression with torsion,
    - biaxial tension-compression with torsion,
    - complete, spatial stress state.
4. Analysed strain states:
    - strain states derived from four above stress states on the basis of the Hooke law.

The strain tensor is derived from the stress tensor on the basis of the Hooke law in order to maintain the fundamental relations between strain tensor components and due to the linear form of the formulae. The current

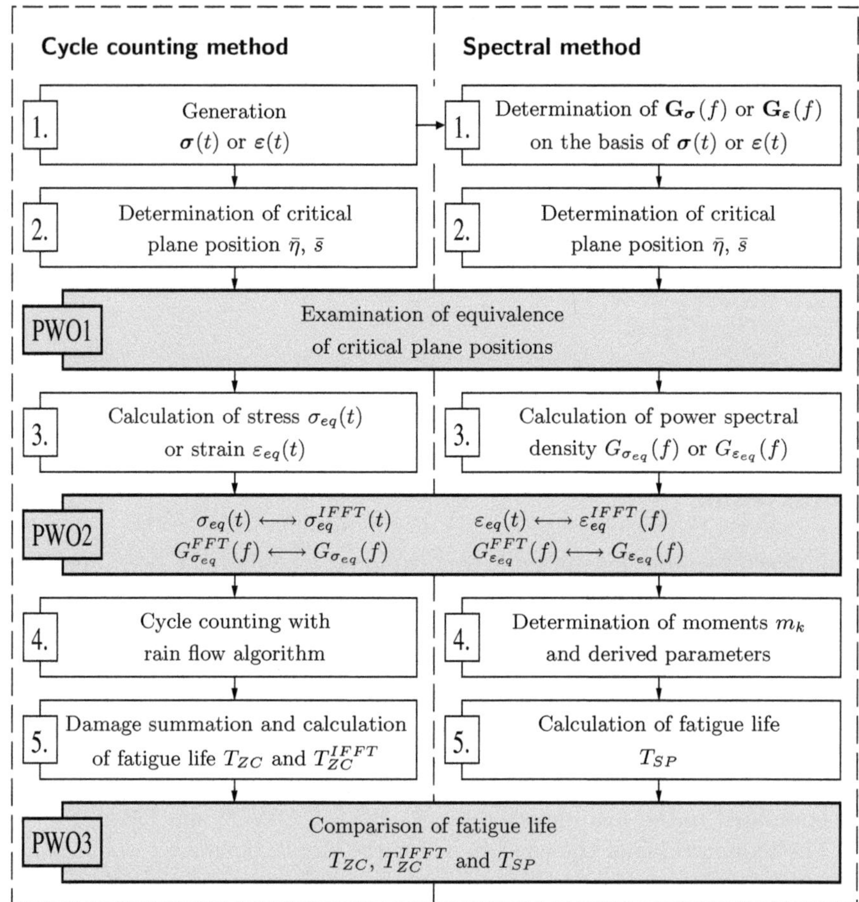

Fig. 5.1. Algorithm of fatigue life calculation applied in simulations including blocks (PWO1 – PWO3) for comparison between cycle counting and spectral methods

authors are familiar with the limitation to the assumptions and consider the resulting values of strain as input quantities for simulations.

The computer simulations apply material constants for steel 18G2A [27], which are summarised in Table 5.1.

Table 5.1. Fatigue and strength characteristics of steel 18G2A

E	ν	σ_{af}	m	N_G	σ'_f	ε'_f	b	c
GPa	—	MPa	—	cycles	MPa	—	—	—
210	0,3	271	7,2	$2{,}735{\cdot}10^6$	782	0,693	−0,118	−0,41

In general, the objectives of the simulations include:

1. Testing of the effect of various strain or stress states on the calculated life time;
2. Comparison of cycle counting method and generalised spectral method on the basis of:
   2.1 calculated positions of critical planes (PWO1),
   2.2 calculated values of equivalent stress or strain and parameters of power spectral density functions of the equivalent quantities (PWO2),
   2.3 calculated fatigue life (PWO3).

5.2 Input Quantities

The input quantities in the simulation calculations include components of stress (3.8) or strain state (3.9) along with material constants (Table 5.1) and the essential strength characteristics. The generation time of random histories $T_o = 682.67$ s (409600 points under sampling frequency 600 Hz) is fixed experimentally. Random histories are generated by means of random number generator with normal probability distribution, which are subsequently filtered with the bandpass Chebyshev Type I filter [89]. The filter order is set experimentally to four and frequency damping outside the to -0.5 dB. Depending on the width of passband the resulting Gaussian histories have narrow-band (passband $f_{min} = 19.5$ Hz and $f_{max} = 20.5$ Hz) and broad-band frequency spectra (passband $f_{min} = 5$ Hz and $f_{max} = 20$ Hz). The output signal obtained from generator was scaled to the variance of values $\mu_x = 1$. Figure 5.2 presents a section of narrow-band frequency history, the probability density distribution function (bar chart) approximated with normal distribution (full line) and power spectral density function. A similar pattern describes a section of broad-band frequency history (Fig. 5.3).

The resulting histories are scaled and included into a vector of stress tensor components $\boldsymbol{\sigma}(t)$, which results in 24 representative stress states. They are marked S1 to S24 and presented in Table 5.2. The corresponding strain states (E1 to E24) are determined in accordance with Hooke law:

$$\varepsilon_{xx}(t) = \frac{1}{E}\left[\sigma_{xx}(t) - \nu\left(\sigma_{yy}(t) + \sigma_{zz}(t)\right)\right],$$

$$\varepsilon_{yy}(t) = \frac{1}{E}\left[\sigma_{yy}(t) - \nu\left(\sigma_{zz}(t) + \sigma_{xx}(t)\right)\right],$$

$$\varepsilon_{zz}(t) = \frac{1}{E}\left[\sigma_{zz}(t) - \nu\left(\sigma_{xx}(t) + \sigma_{yy}(t)\right)\right], \tag{5.1}$$

$$\varepsilon_{xy}(t) = \frac{1+\nu}{E}\sigma_{xy}, \qquad \varepsilon_{xz}(t) = \frac{1+\nu}{E}\sigma_{xz}, \qquad \varepsilon_{yz}(t) = \frac{1+\nu}{E}\sigma_{yz}.$$

A matrix of power spectral density function of stresses $\mathbf{G}_\sigma(f)$ or strains $\mathbf{G}_\varepsilon(f)$ is applied for loading description in the spectral method. In order

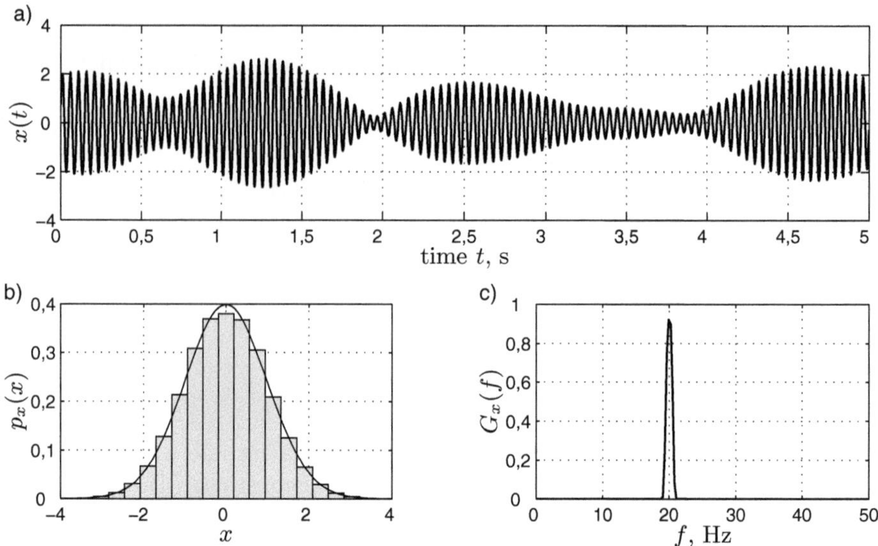

Fig. 5.2. A section of history with narrow-band frequency spectrum (a), including probability density distribution function (b), and power spectral density function (c)

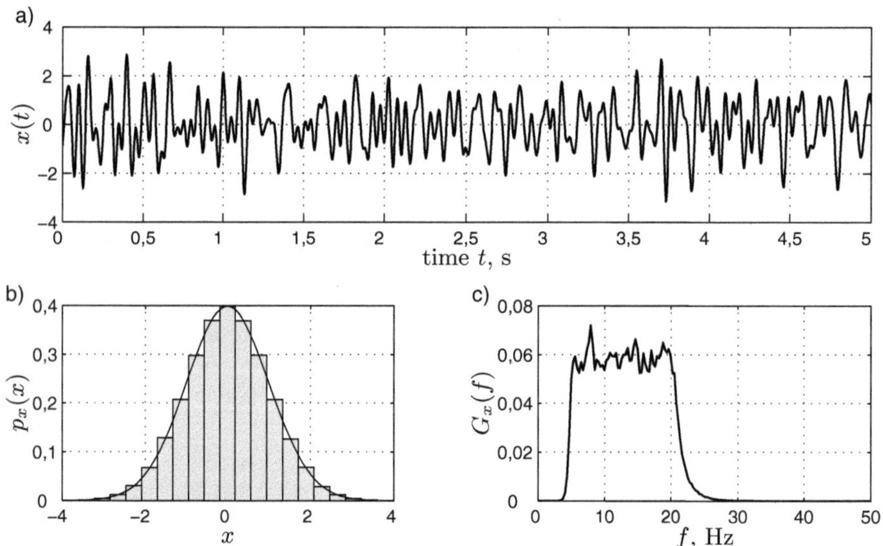

Fig. 5.3. A section of history with broad-band frequency spectrum (a), including probability density distribution function (b), and power spectral density function (c)

to ensure the equivalence of the conditions of loading with regard to spectral method and cycle counting method the matrix of power spectral density functions is determined directly from the vector of stress or strain state component histories. The function `csd` from `MATLAB` [89] is applied for this purpose. It estimates auto- and cross-power spectral density function by means of Welch method. The length of the subsequent signal sections under the analysis is determined by means of Hann type window of the length of 2^{12} points. Each section was subjected to trend and constant component removal. The smooth estimator of power spectral density function resulted from the mean of 100 history sections including total of 100×2^{12} points [6].

Table 5.2: Characteristics of generated stress tensors applied in calculations

Stress state	Covariance $\boldsymbol{\mu}_\sigma$ of tensor components $\boldsymbol{\sigma}(t)$ (MPa^2)	Description
1	2	3
S1 (NB*)	$\begin{bmatrix} 4356 & 4356 & 0 \\ 4356 & 4356 & 0 \\ 0 & 0 & 0 \end{bmatrix}$	$\boldsymbol{\sigma}(t) = [\sigma_{xx}, \sigma_{yy}, 0]$, biaxial tension-compression, complete correlation of components, $r_{xx,yy} = 1$, i.e. proportional loading
S2 (BB*)	$\begin{bmatrix} 4356 & 4356 & 0 \\ 4356 & 4356 & 0 \\ 0 & 0 & 0 \end{bmatrix}$	— „ —
S3	$\begin{bmatrix} 3600 & 3960 & 0 \\ 3960 & 4356 & 0 \\ 0 & 0 & 0 \end{bmatrix}$	$\boldsymbol{\sigma}(t) = [\sigma_{xx}, \sigma_{yy}, 0]$, assymetric biaxial tension-compression, complete correlation of components, $r_{xx,yy} = 1$, σ_{yy} component with larger variance, i.e. proportional loading
S4	$\begin{bmatrix} 3600 & 3960 & 0 \\ 3960 & 4356 & 0 \\ 0 & 0 & 0 \end{bmatrix}$	— „ —
S5	$\begin{bmatrix} 3600 & 16 & 0 \\ 16 & 3600 & 0 \\ 0 & 0 & 0 \end{bmatrix}$	$\boldsymbol{\sigma}(t) = [\sigma_{xx}, \sigma_{yy}, 0]$, biaxial tension-compression, correlation of components $r_{xx,yy} \approx 0$, nonproportional loading
S6	$\begin{bmatrix} 3600 & 49 & 0 \\ 49 & 3600 & 0 \\ 0 & 0 & 0 \end{bmatrix}$	— „ —
S7	$\begin{bmatrix} 3600 & -3600 & 0 \\ -3600 & 3600 & 0 \\ 0 & 0 & 0 \end{bmatrix}$	$\boldsymbol{\sigma}(t) = [\sigma_{xx}, \sigma_{yy}, 0]$, biaxial tension-compression, negative correlation, $r_{xx,yy} = -1$, i.e. proportional loading

Table 5.2: *(continuation)*

1	2	3
S8	$\begin{bmatrix} 3600 & -3600 & 0 \\ -3600 & 3600 & 0 \\ 0 & 0 & 0 \end{bmatrix}$	— ” —
S9	$\begin{bmatrix} 3600 & 0 & 1800 \\ 0 & 0 & 0 \\ 1800 & 0 & 900 \end{bmatrix}$	$\boldsymbol{\sigma}(t) = [\sigma_{xx}, 0, \sigma_{xy}]$, uniaxial tension-compression with torsion, complete correlation of components, $r_{xx,yy} = 1$, i.e. proportional loading
S10	$\begin{bmatrix} 3600 & 0 & 1800 \\ 0 & 0 & 0 \\ 1800 & 0 & 900 \end{bmatrix}$	— ” —
S11	$\begin{bmatrix} 3600 & 0 & -5 \\ 0 & 0 & 0 \\ -5 & 0 & 900 \end{bmatrix}$	$\boldsymbol{\sigma}(t) = [\sigma_{xx}, 0, \sigma_{xy}]$, uniaxial tension-compression with torsion, correlation of components, $r_{xx,xy} \approx 0$, i.e. nonproportional loading
S12	$\begin{bmatrix} 3600 & 0 & 21 \\ 0 & 0 & 0 \\ 21 & 0 & 900 \end{bmatrix}$	— ” —
S13	$\begin{bmatrix} 1764 & 1764 & 882 \\ 1764 & 1764 & 882 \\ 882 & 882 & 441 \end{bmatrix}$	$\boldsymbol{\sigma}(t) = [\sigma_{xx}, \sigma_{yy}, \sigma_{xy}]$, complete plain stress state, complete correlation of components, $r_{xx,yy} = r_{xx,xy} = r_{yy,xy} = 1$, i.e. proportional loading
S14	$\begin{bmatrix} 1764 & 1764 & 882 \\ 1764 & 1764 & 882 \\ 882 & 882 & 441 \end{bmatrix}$	— ” —
S15	$\begin{bmatrix} 3600 & 3600 & 59 \\ 3600 & 3600 & 59 \\ 59 & 59 & 900 \end{bmatrix}$	$\boldsymbol{\sigma}(t) = [\sigma_{xx}, \sigma_{yy}, \sigma_{xy}]$, complete plain stress state, $r_{xx,yy} = 1$, $r_{xx,xy} = r_{yy,xy} \approx 0$, i.e. nonproportional loading

Table 5.2: *(continuation)*

1	2	3
S16	$\begin{bmatrix} 3600 & 3600 & -4 \\ 3600 & 3600 & -4 \\ -4 & -4 & 900 \end{bmatrix}$	— „ —
S17	$\begin{bmatrix} 576 & 576 & 576 & 576 & 576 & 576 \\ 576 & 576 & 576 & 576 & 576 & 576 \\ 576 & 576 & 576 & 576 & 576 & 576 \\ 576 & 576 & 576 & 576 & 576 & 576 \\ 576 & 576 & 576 & 576 & 576 & 576 \\ 576 & 576 & 576 & 576 & 576 & 576 \end{bmatrix}$	$\boldsymbol{\sigma}(t) = [\sigma_{xx}, \sigma_{yy}, \sigma_{zz}, \sigma_{xy}, \sigma_{xz}, \sigma_{yz}]$, spatial stress state, full components correlation, $r = 1$ for every component pair, i.e. proportional loading
S18	$\begin{bmatrix} 576 & 576 & 576 & 576 & 576 & 576 \\ 576 & 576 & 576 & 576 & 576 & 576 \\ 576 & 576 & 576 & 576 & 576 & 576 \\ 576 & 576 & 576 & 576 & 576 & 576 \\ 576 & 576 & 576 & 576 & 576 & 576 \\ 576 & 576 & 576 & 576 & 576 & 576 \end{bmatrix}$	— „ —
S19	$\begin{bmatrix} 2916 & -17 & -3 & -180 & -16 & -41 \\ -17 & 2916 & 40 & -166 & 45 & -172 \\ -3 & 40 & 2916 & -78 & -34 & 85 \\ -180 & -166 & -78 & 2916 & 130 & -6 \\ -16 & 45 & -34 & 130 & 2916 & -17 \\ -41 & -172 & 85 & -6 & -17 & 2916 \end{bmatrix}$	$\boldsymbol{\sigma}(t) = [\sigma_{xx}, \sigma_{yy}, \sigma_{zz}, \sigma_{xy}, \sigma_{xz}, \sigma_{yz}]$, spatial stress state, component correlation, $r_{kl} \approx 0$; $k, l = xx, yy, zz, xy, xz, yz$ for $k \neq l$, i.e. nonproportional loading

Table 5.2: *(continuation)*

1	2	3
S20	$\begin{bmatrix} 2916 & 52 & -67 & 4 & -9 & 22 \\ 52 & 2916 & 59 & 8 & 41 & 4 \\ -67 & 59 & 2916 & 28 & -6 & 59 \\ 4 & 8 & 28 & 2916 & -39 & -9 \\ -9 & 41 & -6 & -39 & 2916 & 39 \\ 22 & 4 & 59 & -9 & 39 & 2916 \end{bmatrix}$	— ,, —
S21	$\begin{bmatrix} 900 & 810 & 720 & 630 & 540 & 450 \\ 810 & 729 & 648 & 567 & 486 & 405 \\ 720 & 648 & 576 & 504 & 432 & 360 \\ 630 & 567 & 504 & 441 & 378 & 315 \\ 540 & 486 & 432 & 378 & 324 & 270 \\ 450 & 405 & 360 & 315 & 270 & 225 \end{bmatrix}$	$\boldsymbol{\sigma}(t) = [\sigma_{xx}, \sigma_{yy}, \sigma_{zz}, \sigma_{xy}, \sigma_{xz}, \sigma_{yz}]$, spatial stress state, full components correlation, $r = 1$ for every component pair, different variance of particular components, i.e. proportional loading
S22	$\begin{bmatrix} 900 & 810 & 720 & 630 & 540 & 450 \\ 810 & 729 & 648 & 567 & 486 & 405 \\ 720 & 648 & 576 & 504 & 432 & 360 \\ 630 & 567 & 504 & 441 & 378 & 315 \\ 540 & 486 & 432 & 378 & 324 & 270 \\ 450 & 405 & 360 & 315 & 270 & 225 \end{bmatrix}$	— ,, —
S23	$\begin{bmatrix} 3600 & 107 & 69 & -19 & -68 & -89 \\ 107 & 2916 & 50 & -6 & -8 & -3 \\ 69 & 50 & 2304 & 52 & -77 & 25 \\ -19 & -6 & 52 & 1764 & -27 & 32 \\ -68 & -8 & -77 & -27 & 1296 & 12 \\ -89 & -3 & 25 & 32 & 12 & 900 \end{bmatrix}$	$\boldsymbol{\sigma}(t) = [\sigma_{xx}, \sigma_{yy}, \sigma_{zz}, \sigma_{xy}, \sigma_{xz}, \sigma_{yz}]$, spatial stress state, component correlation, $r_{kl} \approx 0$; $k, l = xx, yy, zz, xy, xz, yz$ for $k \neq l$, different variance of particular components, i.e. nonproportional loading

Table 5.2: *(continuation)*

1	2	3
S24	$\begin{bmatrix} 3600 & 5 & -4 & -10 & 26 & 12 \\ 5 & 2916 & -11 & -6 & -30 & -7 \\ -4 & -11 & 2304 & -7 & 4 & 2 \\ -10 & -6 & -7 & 1764 & -5 & 1 \\ 26 & -30 & 4 & -5 & 1296 & -8 \\ 12 & -7 & 2 & 1 & -8 & 900 \end{bmatrix}$	— ” —

* symbol for width band frequency spectrum, NB – narrow-band, BB – broad-band
The cases S1, S3, S5, ..., S23 included narrow-band frequency random histories, and the cases S2, S4, S6, ..., S24 – broad-band frequency random histories. Simulation calculations were performed with the **MATLAB** [58] software

5.3 Analysis of Positions of Critical Planes

The awareness of the position of the critical plane, more precisely, the direction cosines of vector normal $\bar{\eta}(l_\eta, m_\eta, n_\eta)$ and vector tangent to the plane $\bar{s}(l_s, m_s, n_s)$ is indispensable for the determination of equivalent stress or strain with application of multiaxial fatigue failure criteria discussed previously. In order to determine the position of the critical plane, the variance method is applied (see Chapter 4.2). The simulations involve the criterion of maximum normal stress in the critical plane (3.21) and maximum normal strain in the critical plane (3.24). In the cycle counting method the covariance matrix is computed directly from random histories in accordance with formula

$$\mu_{x_{kl}} = \mathrm{e}\left[(x_k - \hat{x}_k)(x_l - \hat{x}_l)\right], \qquad (k, l = 1, \ldots, 6), \qquad (5.2)$$

while in the spectral method from the matrices of power spectral density functions in accordance with formula (4.2).

The simulations are applied for determination of the equivalent stress or strain variance for many (20897) possible positions of the critical plane. For the selected multiaxial fatigue failure criteria the awareness of vector normal to plane $\bar{\eta}$ is sufficient for the determination of its position. For the instance above the directional cosines are generated in a way that ensures filling of a half of the sphere describing the tip of the vector normal to the critical plane (Fig. 5.4).

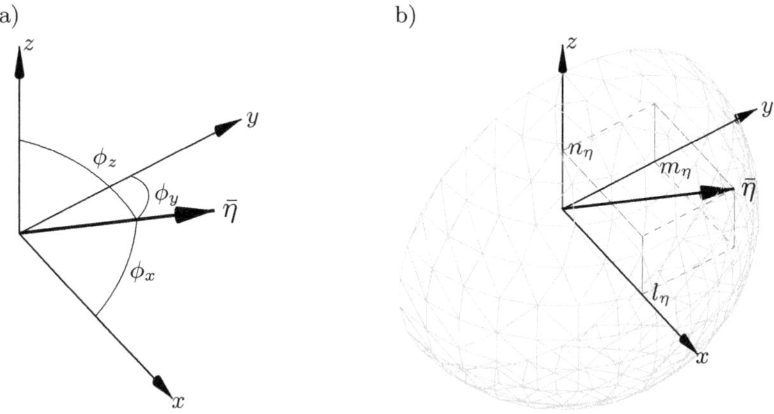

Fig. 5.4. Half of the sphere describing the tip of the vector normal $\bar{\eta}$ to the critical plane in search of maximum variance of equivalent stress or strain

The suitable direction cosines are generated with the limitations resulting from the equation

$$\sqrt{l_\eta^2 + m_\eta^2 + n_\eta^2} = 1. \qquad (5.3)$$

Under the assumption that $l_\eta \in \langle 0; 1 \rangle$ (a half of the sphere, as in Fig. 5.4) it follows that:

$$
\begin{aligned}
m_\eta &\in \left\langle -\sqrt{1 - l_\eta^2}; \ \sqrt{1 - l_\eta^2} \right\rangle , \\
n_\eta &\in \left\langle -\sqrt{1 - l_\eta^2 - m_\eta^2}; \ \sqrt{1 - l_\eta^2 - m_\eta^2} \right\rangle .
\end{aligned}
\tag{5.4}
$$

The relations presented above enable a graphical representation of the variance of the equivalent quantity in the function of the critical plane position determined with two directional cosines of vector normal to the plane. The value of the third direction cosines is determined from formula

$$
l_\eta = \sqrt{1 - m_\eta^2 - n_\eta^2} .
\tag{5.5}
$$

Figures 5.5 to 5.16 present variance charts of equivalent stress and strain, respectively in the function of two directional cosines m_η and n_η (projection of half sphere on a plane). The grey shade marks the value of variance respective for the scale quoted to the right of the particular charts. The triangle marks the locations for which case variance reach maximum values determined in a conventional way in time domain. The square marks the maximum values of variance determined directly from the matrices of power spectral density functions.

It is remarked that maximum variance for plain stress state (cases S1 to S16) occurs only when the critical plane coincides with X-Y plane, i.e. for $\phi_z = \dfrac{\pi}{2}$. A similar distribution of maximums is distinguished for strain (cases E1 to E16), although the cases involve spatial strain states.

In several cases the algorithm of seeking the maximum of a variance provides more than single solution. The justification of the phenomenon can be twofold. The first is easily discernible in Fig. 5.5 and Fig. 5.11. The reason for the phenomenon is associated with constant variance of equivalent quantity as a result of the selection of components of stress tensor and its values. For instance, during loading under biaxial tension-compression with component correlation $r_{xx,yy} = 1$ (cases S1 and S2) the variance method indicates that the occurrence of the position of the critical plane for every $m_\eta \in \langle -1, 1 \rangle$ with equal probability, while $n_\eta = 0$ and $l_\eta = \sqrt{1 - m_\eta^2}$. Following the analysis of mathematical forms of the applied multiaxial fatigue failure criteria the other justification for the occurrence of several variance extremes $\mu_{\sigma_{eq}}$ and $\mu_{\varepsilon_{eq}}$ could be identified. In several cases (e.g. S7, S8, E3, E4) under the selection of suitable variances of stress or strain tensor components, the functions $\mu_{\sigma_{eq}}$ and $\mu_{\varepsilon_{eq}}$ demonstrate periodical qualities. As a result of equal variance in the positions it could be presumed that fatigue failure initiate in one of the determined planes. The selection of the plane is random and depends on the quality of material.

It is noticeable that the shapes of the determined functions $\mu_{\sigma_{eq}}$ and $\mu_{\varepsilon_{eq}}$ in time domain coincide with the shapes of adequate functions determined by means of spectral method. Apart form that, the maximum values are equal. Differences are observed between the spectral method and cycle counting method only in several cases. They result from numerical errors in the course of covariance matrix calculation by the methods. It is particularly discernible in the cases when tensor components have similar variances (S20, S24). In the cases the small differences during estimation of covariance matrixes with the two methods result in large difference of the estimated positions of critical planes.

Table 5.3 summarises the direction cosines of vector normal to the critical plane for which the variance of equivalent stress or strain reach the maximum. For the purpose of the simplification of data analysis the angles of vector normal $\bar{\eta}$ with co-ordinate axes x, y, z are defined.

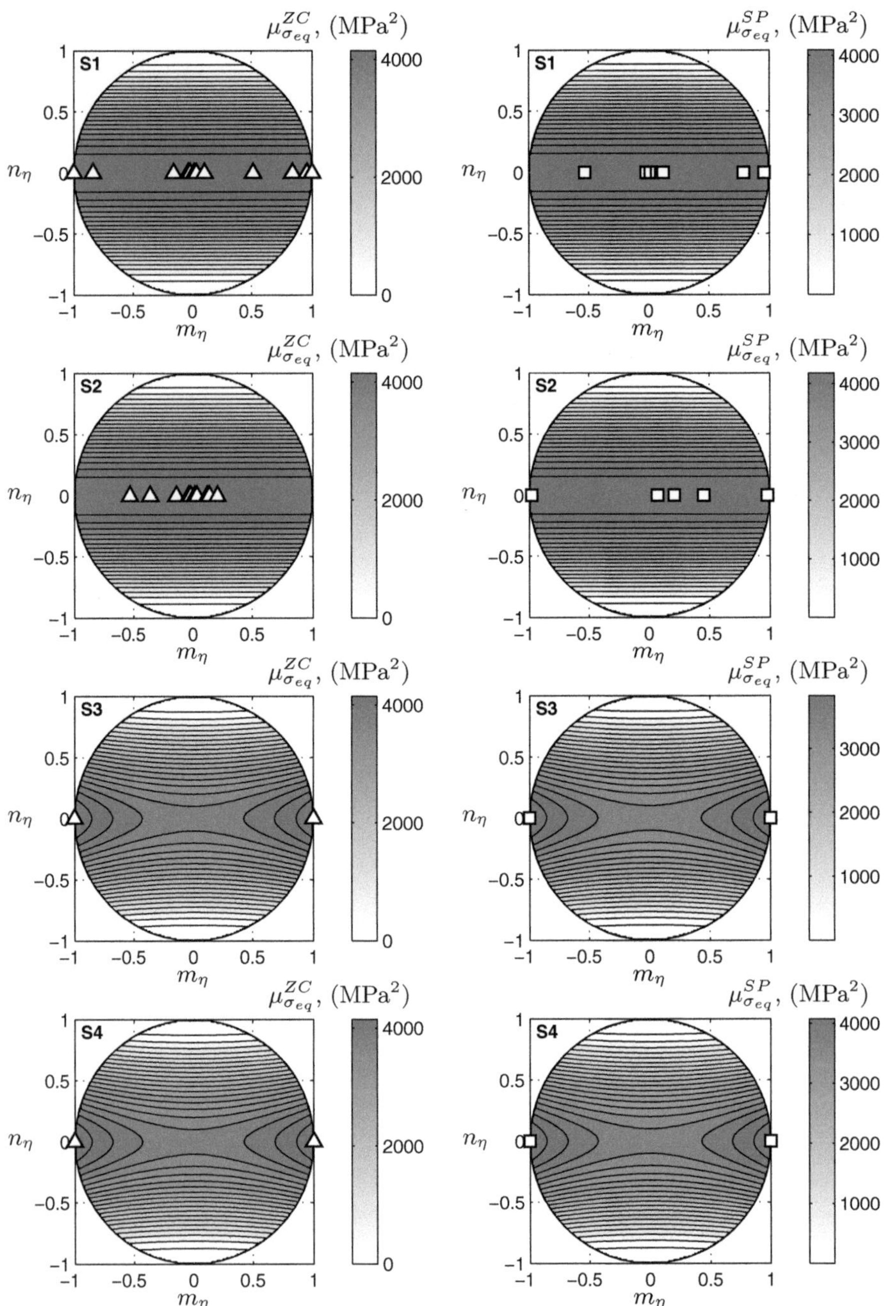

Fig. 5.5. Variance charts of equivalent stress $\mu_{\sigma_{eq}}(m_\eta, n_\eta)$ for conducted simulations. The cases of loading defined by stress tensors (S1–S4)

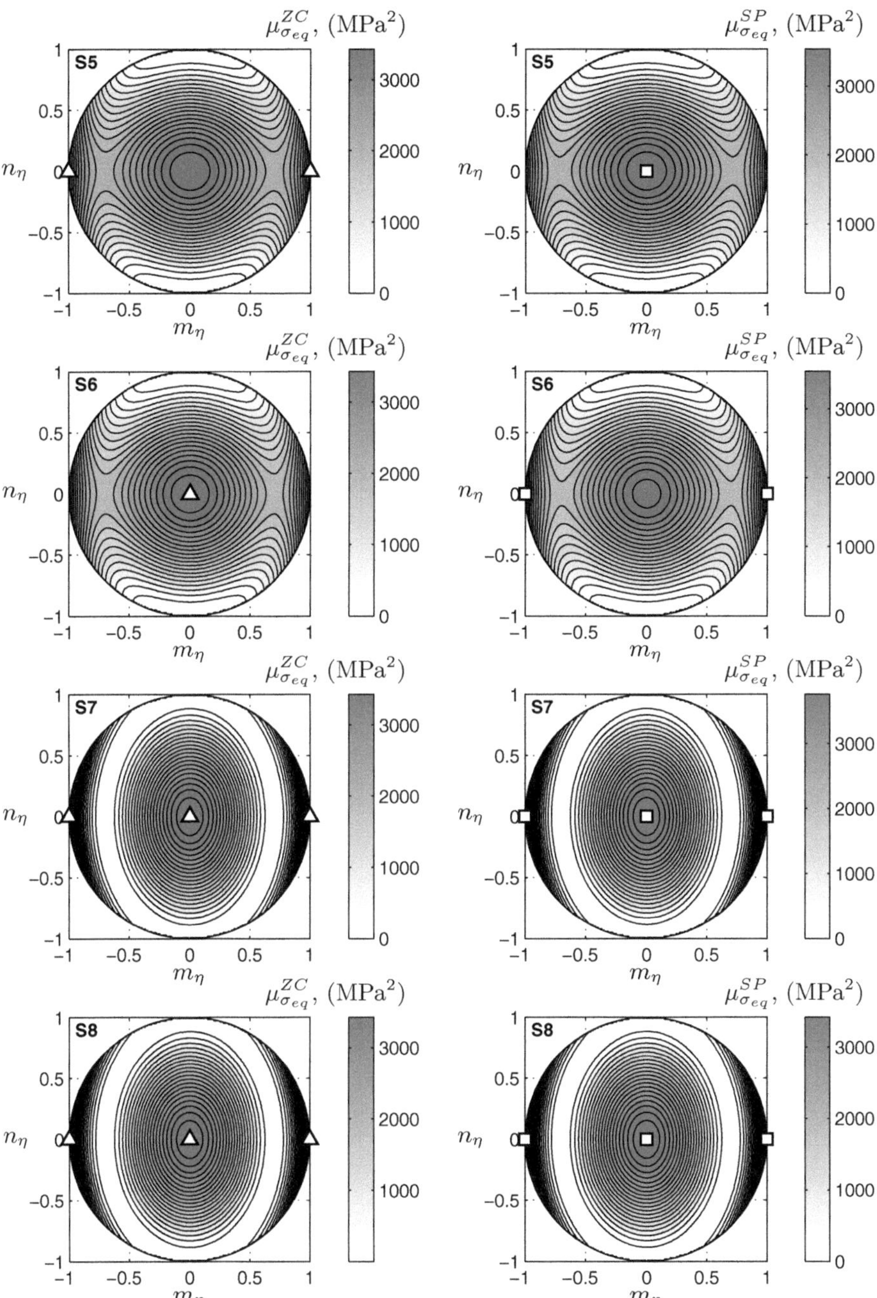

Fig. 5.6. Variance charts of equivalent stress $\mu_{\sigma_{eq}}(m_\eta, n_\eta)$ for conducted simulations. The cases of loading defined by stress tensors (S5–S8)

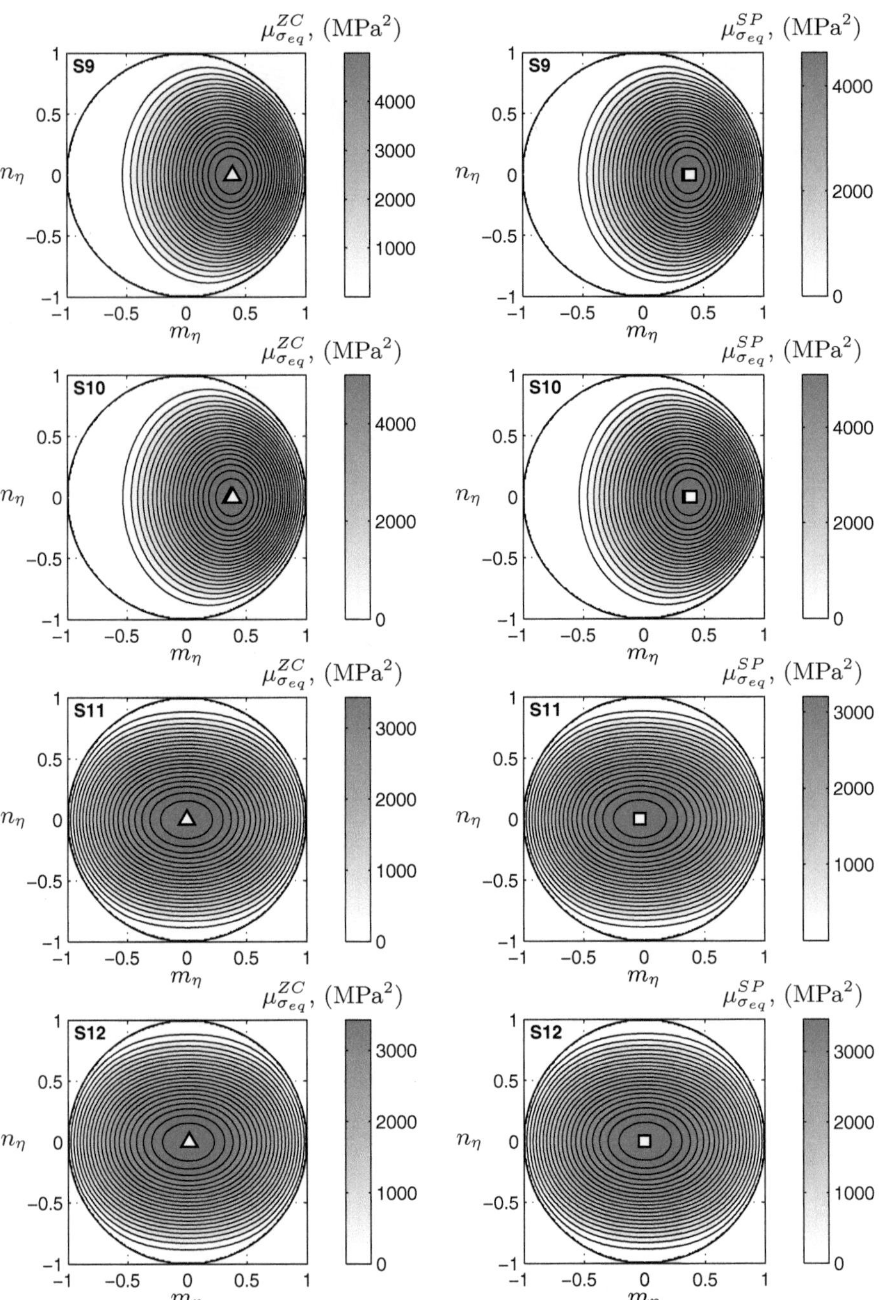

Fig. 5.7. Variance charts of equivalent stress $\mu_{\sigma_{eq}}(m_\eta, n_\eta)$ for conducted simulations. The cases of loading defined by stress tensors (S9–S12)

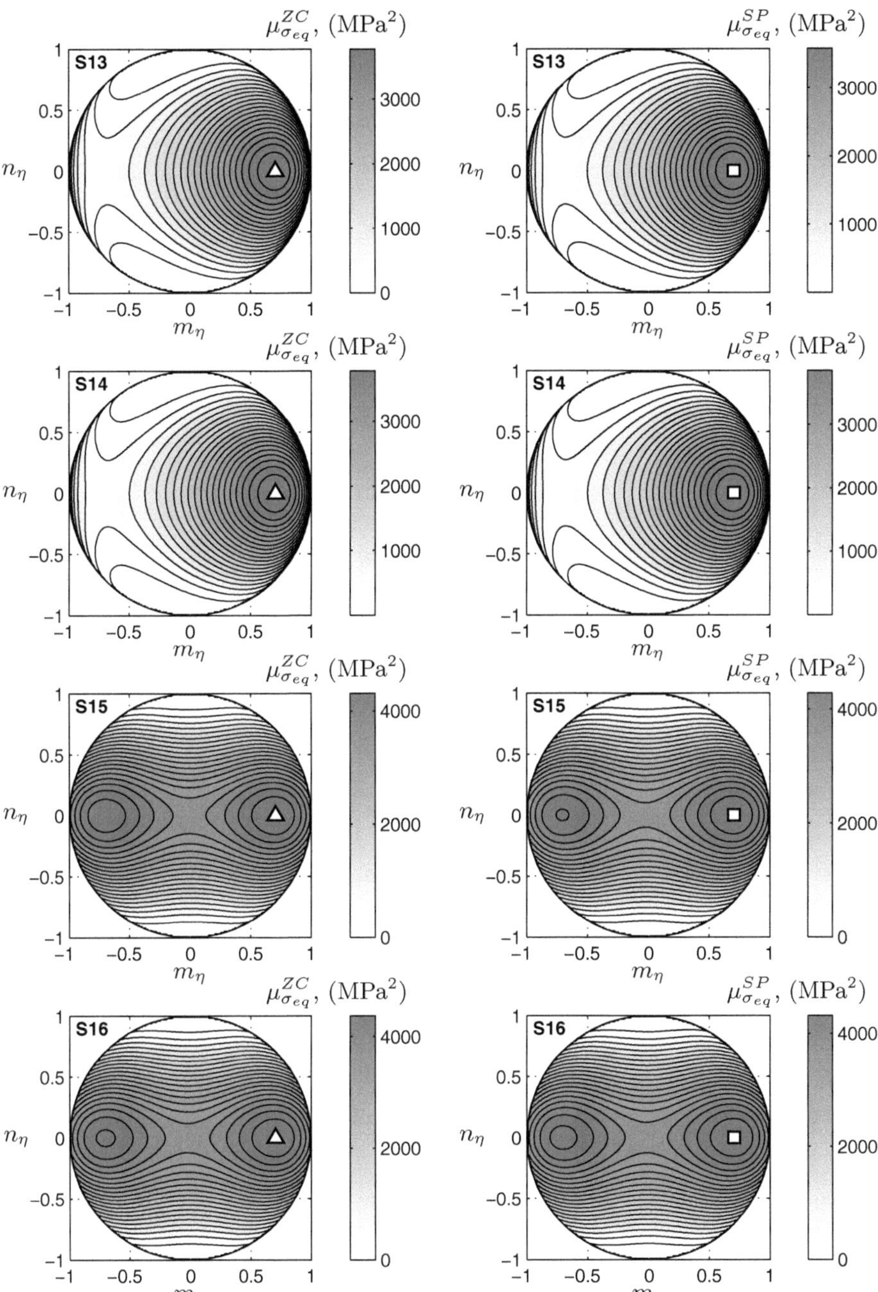

Fig. 5.8. Variance charts of equivalent stress $\mu_{\sigma_{eq}}(m_\eta, n_\eta)$ for conducted simulations. The cases of loading defined by stress tensors (S13–S16)

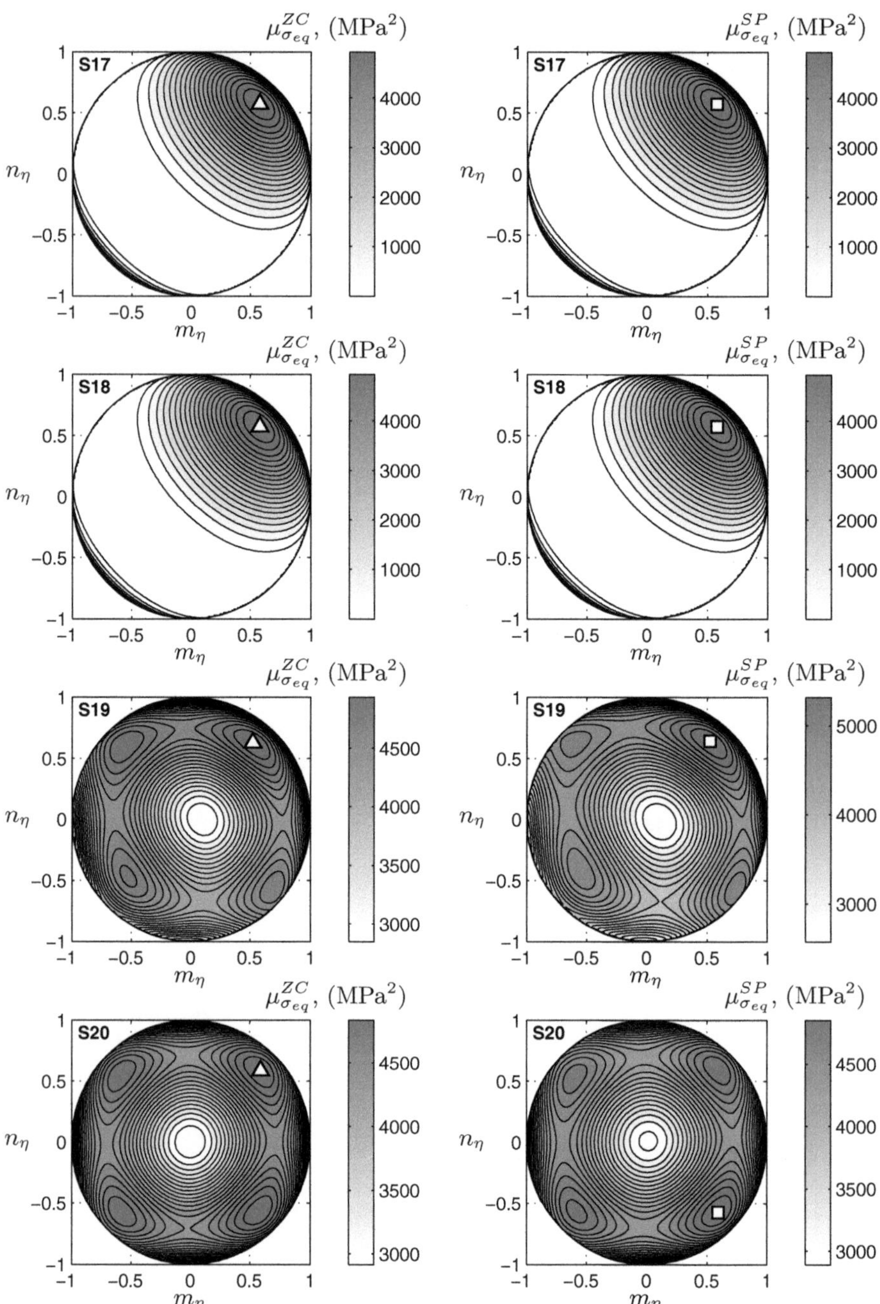

Fig. 5.9. Variance charts of equivalent stress $\mu_{\sigma_{eq}}(m_\eta, n_\eta)$ for conducted simulations. The cases of loading defined by stress tensors (S17–S20)

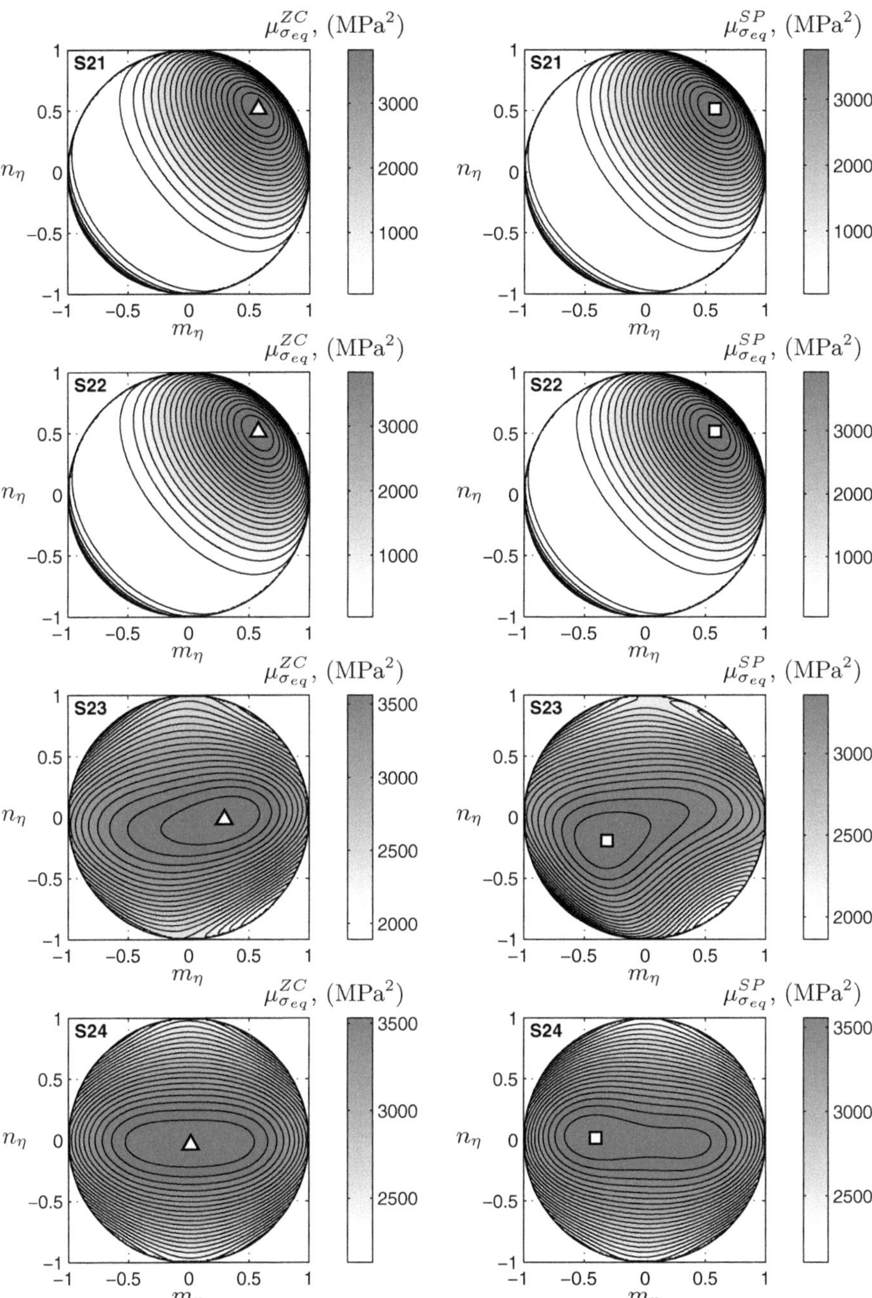

Fig. 5.10. Variance charts of equivalent stress $\mu_{\sigma_{eq}}(m_\eta, n_\eta)$ for conducted simulations. The cases of loading defined by stress tensors (S21–S24)

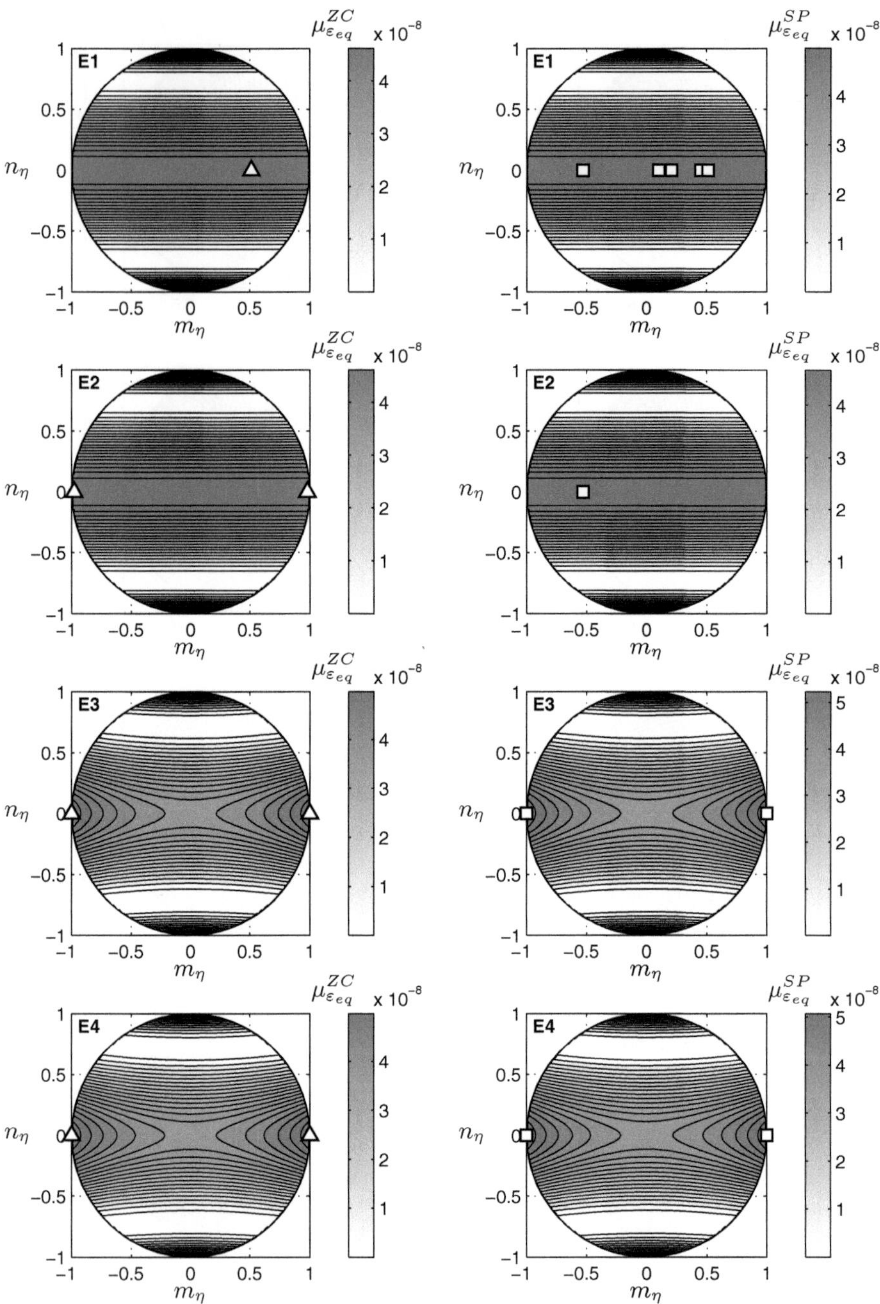

Fig. 5.11. Variance charts of equivalent strain $\mu_{\varepsilon_{eq}}(m_\eta, n_\eta)$ for conducted simulations. The cases of loading defined by strain tensors (E1–E4)

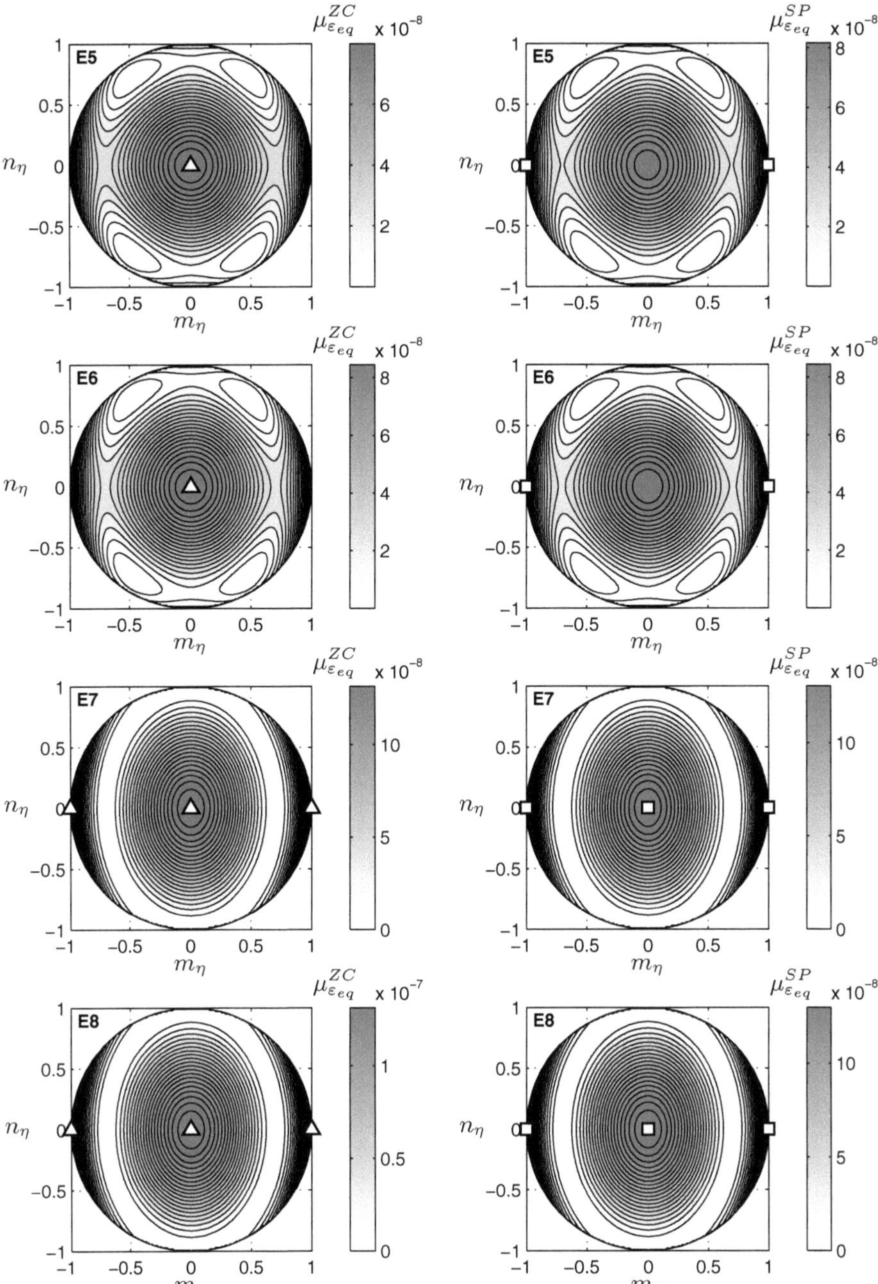

Fig. 5.12. Variance charts of equivalent strain $\mu_{\varepsilon_{eq}}(m_\eta, n_\eta)$ for conducted simulations. The cases of loading defined by strain tensors (E5–E8)

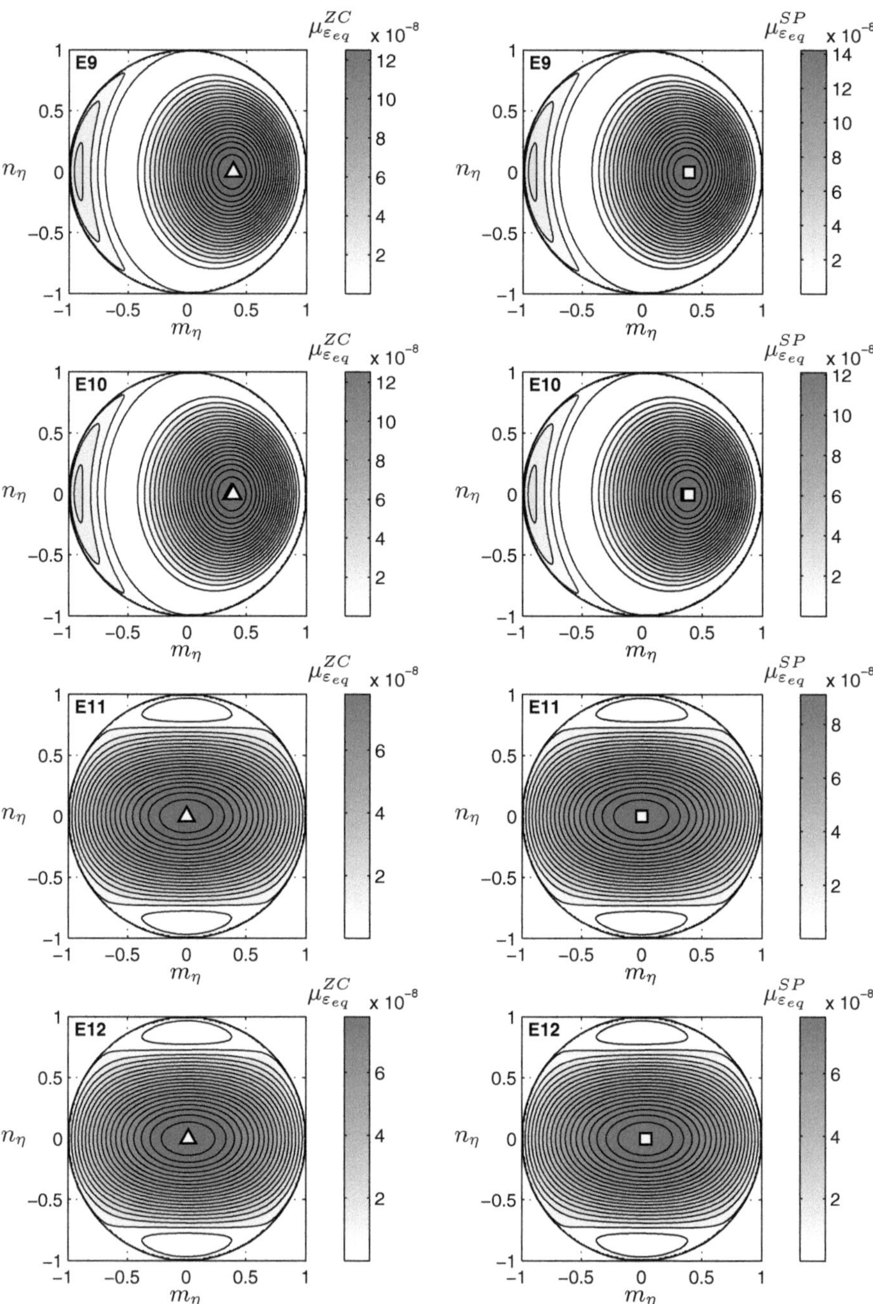

Fig. 5.13. Variance charts of equivalent strain $\mu_{\varepsilon_{eq}}(m_\eta, n_\eta)$ for conducted simulations. The cases of loading defined by strain tensors (E9–E12)

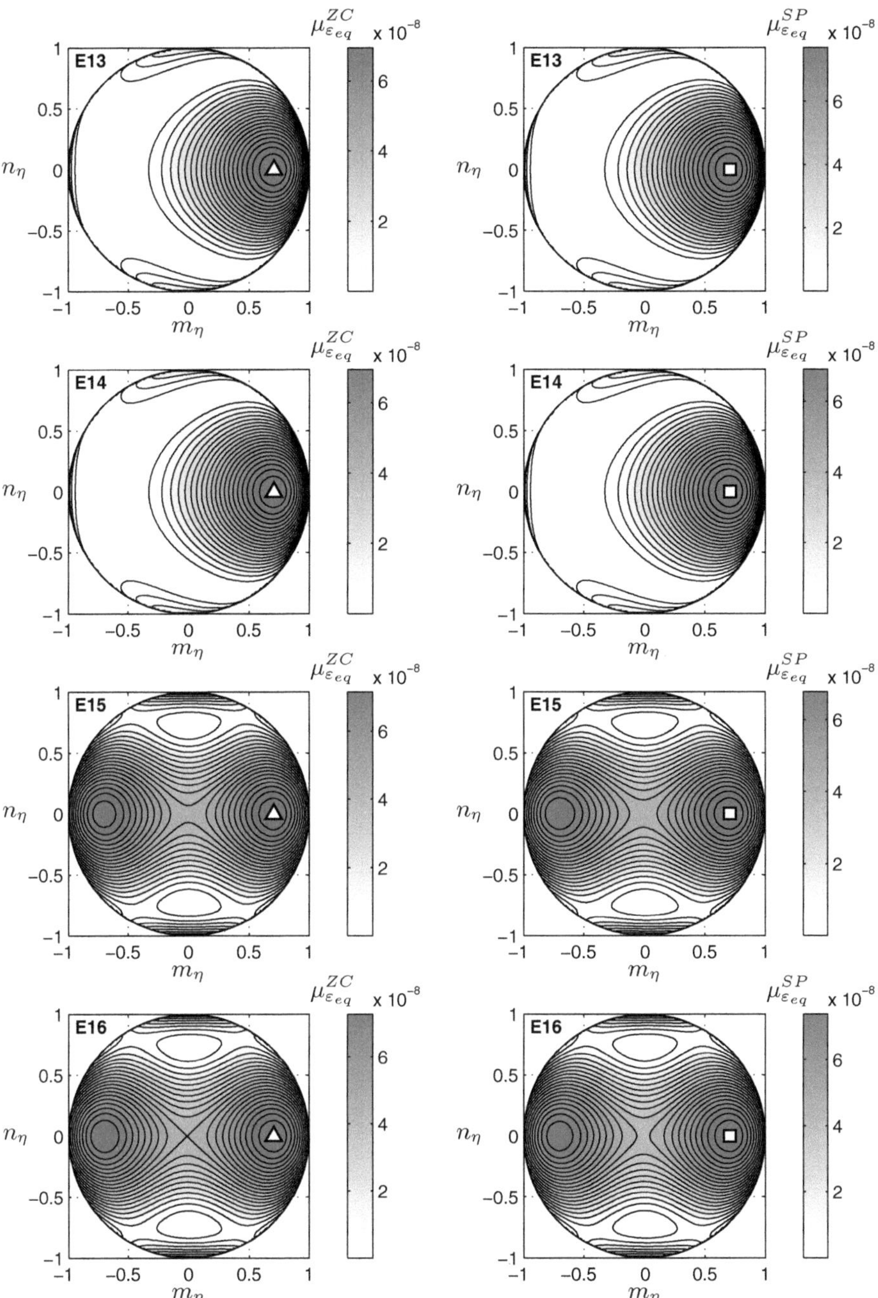

Fig. 5.14. Variance charts of equivalent strain $\mu_{\varepsilon eq}(m_\eta, n_\eta)$ for conducted simulations. The cases of loading defined by strain tensors (E13–E16)

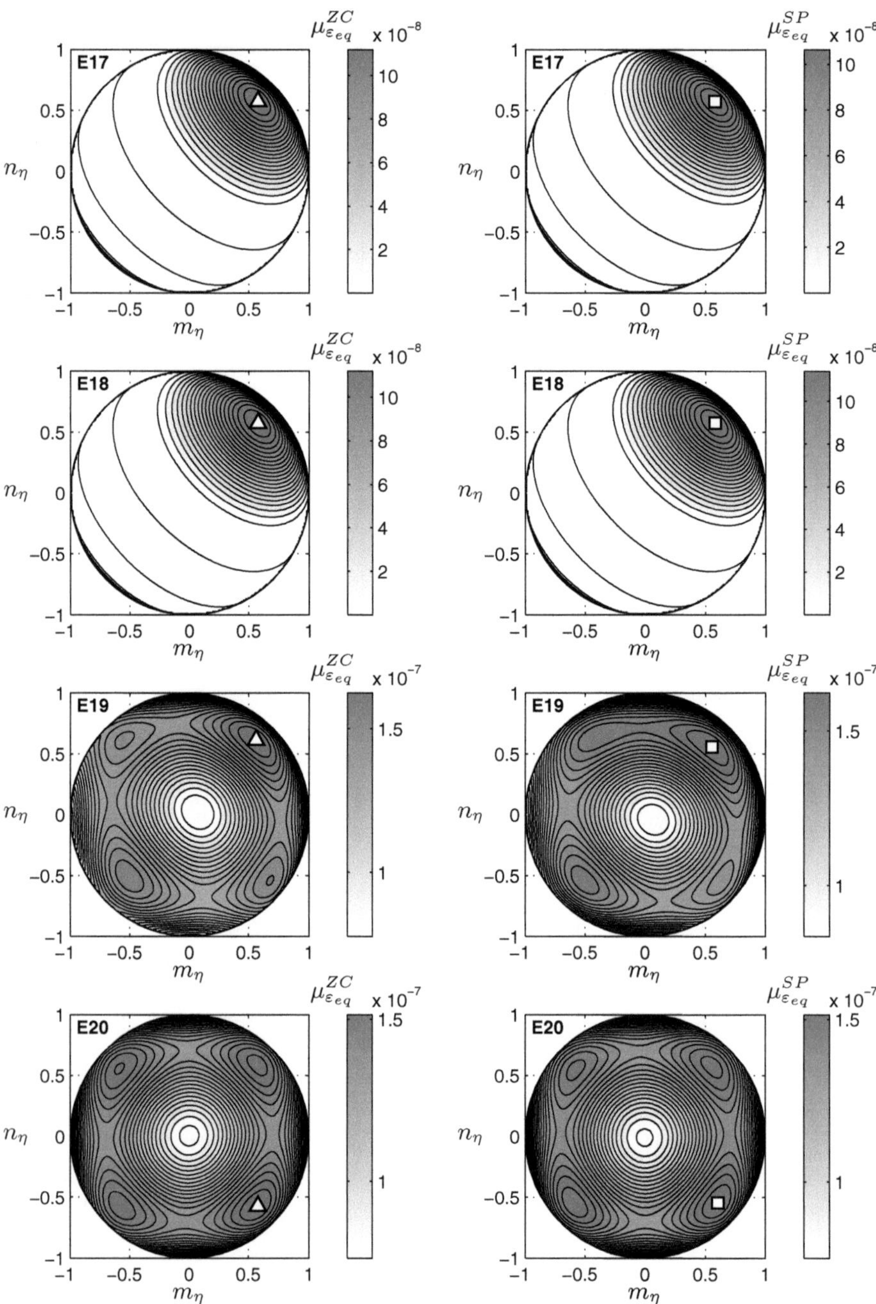

Fig. 5.15. Variance charts of equivalent strain $\mu_{\varepsilon_{eq}}(m_\eta, n_\eta)$ for conducted simulations. The cases of loading defined by strain tensors (E17–E20)

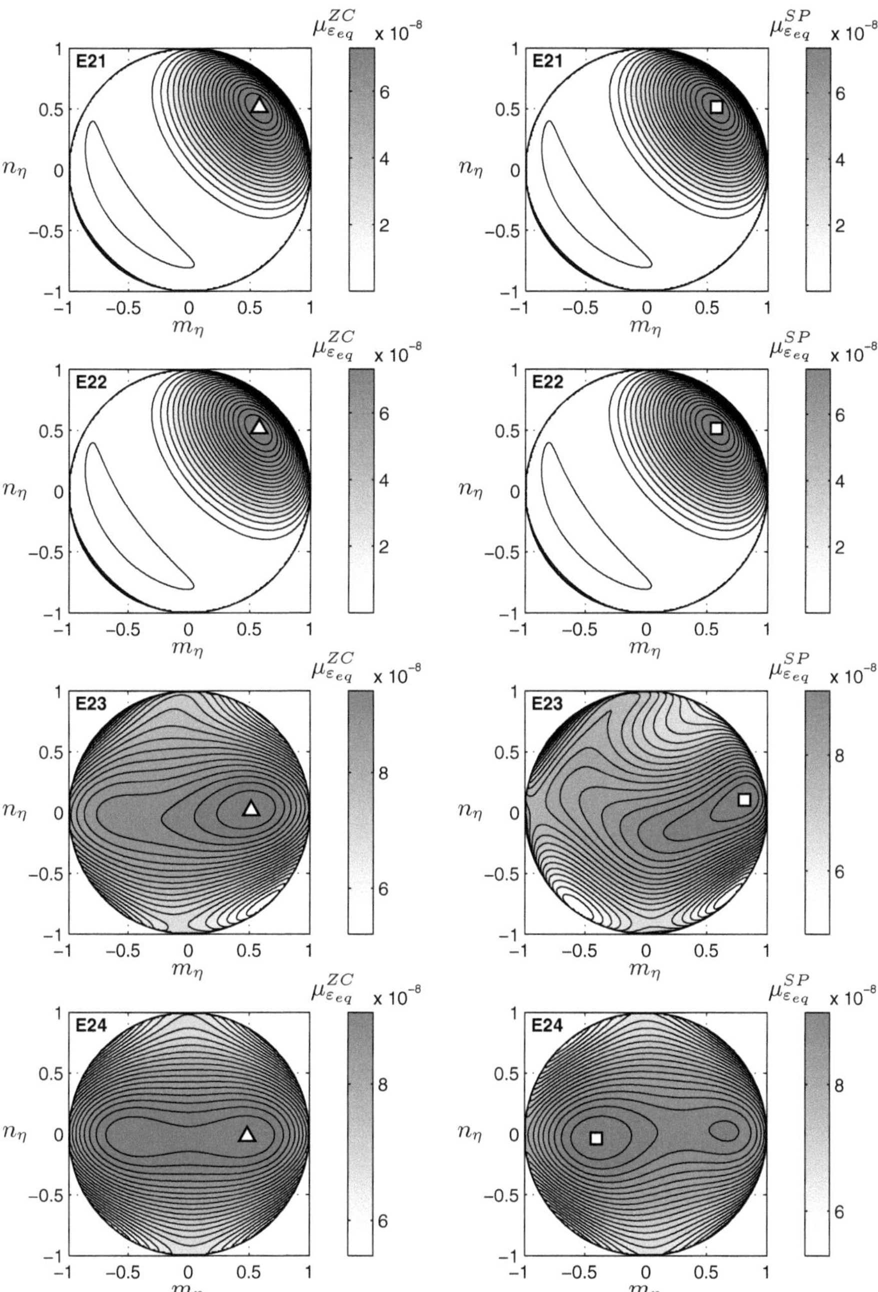

Fig. 5.16. Variance charts of equivalent strain $\mu_{\varepsilon_{eq}}(m_\eta, n_\eta)$ for conducted simulations. The cases of loading defined by strain tensors (E21–E24)

Table 5.3: Directional cosines of normal vector $\bar{\eta}(l_\eta, m_\eta, n_\eta)$ to critical plane determined by the maximum variance method

Case of loading	$[l_\eta, m_\eta, n_\eta]_{ZC}$ —	$[\phi_x, \phi_y, \phi_z]_{ZC}$ degree	$[l_\eta, m_\eta, n_\eta]_{SP}$ —	$[\phi_x, \phi_y, \phi_z]_{SP}$ degree
1	2	3	4	5
S1	[0.035 -0.999 0]	[88 178 90]	[0.848 -0.53 0]	[32 122 90]
	[0.545 -0.839 0]	[57 147 90]	[1 -0.017 0]	[1 91 90]
	[0.988 -0.156 0]	[9 99 90]	[1 0.017 0]	[1 89 90]
	[0.999 -0.035 0]	[2 92 90]	[0.998 0.07 0]	[4 86 90]
	[1 -0.017 0]	[1 91 90]	[0.995 0.105 0]	[6 84 90]
	[1 0.017 0]	[1 89 90]	[0.993 0.122 0]	[7 83 90]
	[0.999 0.035 0]	[2 88 90]	[0.616 0.788 0]	[52 38 90]
	[0.995 0.105 0]	[6 84 90]	[0.292 0.956 0]	[73 17 90]
	[0.857 0.515 0]	[31 59 90]		
	[0.545 0.839 0]	[57 33 90]		
	[0.292 0.956 0]	[73 17 90]		
	[0.035 0.999 0]	[88 2 90]		
E1	[0.857 0.515 0]	[31 59 90]	[0.848 -0.53 0]	[32 122 90]
			[0.995 0.105 0]	[6 84 90]
			[0.978 0.208 0]	[12 78 90]
			[0.891 0.454 0]	[27 63 90]
			[0.857 0.515 0]	[31 59 90]
S2	[0.848 -0.53 0]	[32 122 90]	[0.208 -0.978 0]	[78 168 90]
	[0.934 -0.358 0]	[21 111 90]	[0.998 0.07 0]	[4 86 90]
	[0.99 -0.139 0]	[8 98 90]	[0.978 0.208 0]	[12 78 90]
	[0.999 -0.035 0]	[2 92 90]	[0.891 0.454 0]	[27 63 90]
	[1 -0.017 0]	[1 91 90]	[0.208 0.978 0]	[78 12 90]
	[1 0.017 0]	[1 89 90]		

Table 5.3: *(continuation)*

1	2	3	4	5
	[0.999 0.035 0]	[2 88 90]		
	[0.993 0.122 0]	[7 83 90]		
	[0.99 0.139 0]	[8 82 90]		
	[0.978 0.208 0]	[12 78 90]		
E2	[0.208 −0.978 0]	[78 168 90]	[0.848 −0.53 0]	[32 122 90]
	[0.208 0.978 0]	[78 12 90]		
S3	[0 −1 0]	[90 180 90]	[0 −1 0]	[90 180 90]
	[0 1 0]	[90 0 90]	[0 1 0]	[90 0 90]
E3	[0 −1 0]	[90 180 90]	[0 −1 0]	[90 180 90]
	[0 1 0]	[90 0 90]	[0 1 0]	[90 0 90]
S4	[0 −1 0]	[90 180 90]	[0 −1 0]	[90 180 90]
	[0 1 0]	[90 0 90]	[0 1 0]	[90 0 90]
E4	[0 −1 0]	[90 180 90]	[0 −1 0]	[90 180 90]
	[0 1 0]	[90 0 90]	[0 1 0]	[90 0 90]
S5	[0 −1 0]	[90 180 90]	[1 0 0]	[0 90 90]
	[0 1 0]	[90 0 90]		
E5	[1 0 0]	[0 90 90]	[0 −1 0]	[90 180 90]
			[0 1 0]	[90 0 90]
S6	[1 0 0]	[0 90 90]	[0 −1 0]	[90 180 90]
			[0 1 0]	[90 0 90]
E6	[1 0 0]	[0 90 90]	[0 −1 0]	[90 180 90]
			[0 1 0]	[90 0 90]
S7	[0 −1 0]	[90 180 90]	[0 −1 0]	[90 180 90]
	[1 0 0]	[0 90 90]	[1 0 0]	[0 90 90]
	[0 1 0]	[90 0 90]	[0 1 0]	[90 0 90]

Table 5.3: *(continuation)*

1	2	3	4	5
E7	[0 -1 0]	[90 180 90]	[0 -1 0]	[90 180 90]
	[1 0 0]	[0 90 90]	[1 0 0]	[0 90 90]
	[0 1 0]	[90 0 90]	[0 1 0]	[90 0 90]
S8	[0 -1 0]	[90 180 90]	[0 -1 0]	[90 180 90]
	[1 0 0]	[0 90 90]	[1 0 0]	[0 90 90]
	[0 1 0]	[90 0 90]	[0 1 0]	[90 0 90]
E8	[0 -1 0]	[90 180 90]	[0 -1 0]	[90 180 90]
	[1 0 0]	[0 90 90]	[1 0 0]	[0 90 90]
	[0 1 0]	[90 0 90]	[0 1 0]	[90 0 90]
S9	[0.921 0.391 0]	[23 67 90]	[0.927 0.375 0]	[22 68 90]
			[0.921 0.391 0]	[23 67 90]
E9	[0.921 0.391 0]	[23 67 90]	[0.921 0.391 0]	[23 67 90]
S10	[0.927 0.375 0]	[22 68 90]	[0.927 0.375 0]	[22 68 90]
	[0.921 0.391 0]	[23 67 90]	[0.921 0.391 0]	[23 67 90]
E10	[0.927 0.375 0]	[22 68 90]	[0.927 0.375 0]	[22 68 90]
	[0.921 0.391 0]	[23 67 90]	[0.921 0.391 0]	[23 67 90]
S11	[1 0 0]	[0 90 90]	[0.999 -0.035 0]	[2 92 90]
E11	[1 0 0]	[0 90 90]	[1 0 0]	[0 90 90]
S12	[1 0.017 0]	[1 89 90]	[1 0 0]	[0 90 90]
E12	[1 0.017 0]	[1 89 90]	[0.999 0.035 0]	[2 88 90]
S13	[0.707 0.707 0]	[45 45 90]	[0.707 0.707 0]	[45 45 90]
E13	[0.707 0.707 0]	[45 45 90]	[0.707 0.707 0]	[45 45 90]
S14	[0.707 0.707 0]	[45 45 90]	[0.707 0.707 0]	[45 45 90]
E14	[0.707 0.707 0]	[45 45 90]	[0.707 0.707 0]	[45 45 90]
S15	[0.707 0.707 0]	[45 45 90]	[0.707 0.707 0]	[45 45 90]

Table 5.3: *(continuation)*

1	2	3	4	5
E15	[0.707 0.707 0]	[45 45 90]	[0.707 0.707 0]	[45 45 90]
S16	[0.707 0.707 0]	[45 45 90]	[0.707 0.707 0]	[45 45 90]
E16	[0.707 0.707 0]	[45 45 90]	[0.707 0.707 0]	[45 45 90]
S17	[0.579 0.579 0.574]	[55 55 55]	[0.579 0.579 0.574]	[55 55 55]
E17	[0.579 0.579 0.574]	[55 55 55]	[0.579 0.579 0.574]	[55 55 55]
S18	[0.579 0.579 0.574]	[55 55 55]	[0.579 0.579 0.574]	[55 55 55]
E18	[0.579 0.579 0.574]	[55 55 55]	[0.579 0.579 0.574]	[55 55 55]
S19	[0.574 0.524 0.629]	[55 58 51]	[0.56 0.523 0.643]	[56 58 50]
E19	[0.551 0.563 0.616]	[57 56 52]	[0.616 0.555 0.559]	[52 56 56]
S20	[0.553 0.59 0.588]	[56 54 54]	[0.567 0.591 −0.574]	[55 54 125]
E20	[0.579 0.579 −0.574]	[55 55 125]	[0.584 0.602 −0.545]	[54 53 123]
S21	[0.633 0.578 0.515]	[51 55 59]	[0.633 0.578 0.515]	[51 55 59]
E21	[0.633 0.578 0.515]	[51 55 59]	[0.633 0.578 0.515]	[51 55 59]
S22	[0.633 0.578 0.515]	[51 55 59]	[0.633 0.578 0.515]	[51 55 59]
E22	[0.633 0.578 0.515]	[51 55 59]	[0.633 0.578 0.515]	[51 55 59]
S23	[0.956 0.292 −0.017]	[17 73 91]	[0.929 −0.317 −0.191]	[22 108 101]
E23	[0.857 0.515 0.017]	[31 59 89]	[0.57 0.815 0.105]	[55 35 84]
S24	[0.999 0.017 −0.035]	[2 89 92]	[0.913 −0.407 0.017]	[24 114 89]
E24	[0.874 0.485 −0.017]	[29 61 91]	[0.913 −0.406 −0.035]	[24 114 92]

5.4 Analysis of Equivalent Quantities

Block 3 of algorithm for fatigue life calculation by means of the cycle counting method (Fig. 5.1) is consists in the computation of histories of equivalent stress $\sigma_{eq}(t)$ or strain $\varepsilon_{eq}(t)$, whereas the Block 3 of algorithm of the spectral method, the power spectral density $G_{\sigma_{eq}}(f)$ or $G_{\varepsilon_{eq}}(f)$ is computed. They are determined by application of multiaxial fatigue failure criteria in time and frequency domains, respectively. A direct comparison between the quantities is impossible due to the differences between them. Therefore, random history $\sigma_{eq}^{FFT}(t)$ is derived from power spectral density $G_{eq}(f)$ and power spectral density $G_{eq}^{FFT}(t)$ is derived from equivalent stress $\sigma_{eq}(t)$. Strain history is treated in an analogous manner. The resulting values are subsequently compared. In order to present the sequence of calculation leading to the results, block PWO2 (diagram in Fig. 5.1) is elaborated. A unique case involving stress history in Fig. 5.17 is presented.

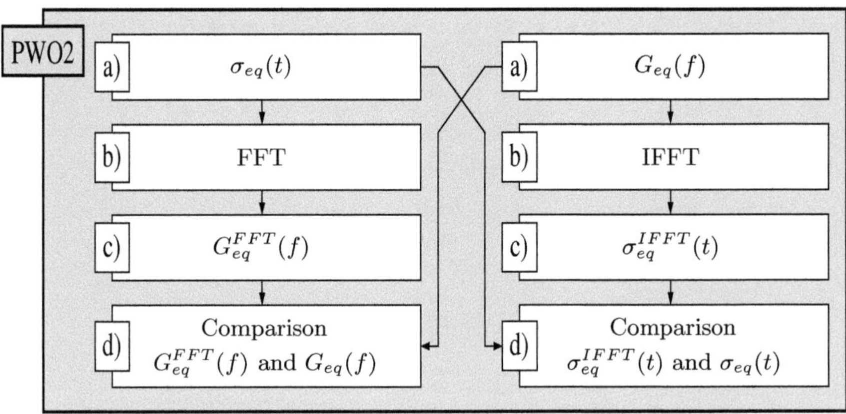

Fig. 5.17. Algorithm of comparison of results of equivalent stress history calculation in time domain and power spectral density in frequency domain

5.4.1 Random Histories

$\sigma_{eq}^{IFFT}(t)$ history is computed with the inverse fast Fourier transform (IFFT) method. It consists in the application of algorithm of inverse fast Fourier transform for the generation of random histories on the basis of amplitudes of component harmonics and their phase shifts. The amplitudes of component harmonics A_i are derived directly from the estimator of power spectral density function,

$$A_i = \sqrt{2G_{eq_i}\,\Delta f}\,, \tag{5.6}$$

where: A_i – amplitude of i-th harmonic component,
   G_{eq_i} – i-th value of estimator of power spectral density function,
   Δf – distance between the successive discrete values of estimator of power spectral density – frequency interval,

while the phase shift is determined by the generation of random numbers with uniform distribution in the range $\langle 0, \ldots, 2\pi \rangle$. The bibliography includes other methods of random history generation on the basis of power spectral density [70, 77]. The current method is selected due to the opposition to power spectral density determination by fast Fourier transform (FFT). The only drawback of the method is the absence the phase shift of component harmonics, which could not be retrieved while they are required during generation.

A comparison of random histories $\sigma_{eq}(t)$ and $\sigma_{eq}^{IFFT}(t)$ is carried out in several stages. At first both random histories are examined for equivalence of probability density functions. For the reason the Kolmogorov-Smirnov [96] significance test of two random variables is performed for probability density function under the assumption of the significance level $\alpha_{KS} = 0.05$. The maximum of the difference of distribution function of the analysed variables is tested. It could be remarked that the test is very sensitive, as it is sufficient to increase the value of one of the tested variables by 5% for the negative verification of the hypothesis of equality of probability distributions. The test results prove the lack of premises for the rejection of the hypothesis concerning the equivalence of distributions in the analysed cases.

Subsequently, a comparison is made for the amplitude distributions of random histories $\sigma_{eq}(t)$ and $\sigma_{eq}^{IFFT}(t)$ obtained from the rain flow algorithm [1]. Amplitudes from the random histories are compared, and the Kolmogorov-Smirnov test [96] for the equivalence of distribution functions is performed just as in the previous instance. The testing indicates that there are no reasons for the rejection of the hypothesis of the equivalence of both amplitude probability distributions.

5.4.2 Power Spectral Densities

Power spectral density functions are computed from the history of equivalent stress by FFT algorithm in a similar manner as from the matrices of power spectral density of stress and strain states components (Block 1 of simulation algorithm, Fig. 5.1). A comparison is made between the moments of spectral density m_k, as they constitute the foundation for the determination of statistical parameters applied in spectral formulae for the calculations of fatigue life. In the course of calculations, identical values of power spectral density estimators of equivalent quantities are obtained in all cases. It could have been anticipated as the determination of the value of equivalent quantity in time or frequency domain is equivalent from the theoretical point of view. The comparison additionally indicates sufficient accuracy of numerical calculations and rejected the possibility of essential error in the further analysis.

5.5 Analysis of Fatigue Life

Block PWO3 of the algorithm involves the comparison of the results of the calculated fatigue life. Symbols T_{ZC} and T_{ZC}^{IFFT} mark the fatigue life calculated with the cycle counting method, with a difference that for the case of the latter one the method of the inverse fast Fourier transform is applied for the generation of the equivalent stress and strain histories (the histories registered in block PWO2 are applied). In the cycle counting method the Palmgen-Miner linear hypothesis of damage accumulation is applied

$$T_{ZC} = \frac{T_o}{D(T_o)} = \frac{T_o}{\sum_k \dfrac{n_k}{N_k}} , \tag{5.7}$$

where: T_o – observation time, loading history under cycle counting,
       $D(T_o)$ – damage induced by loading during time T_o,
       k – number of levels of loading amplitude,
       n_k – number of cycles of amplitudes σ_{ak} or ε_{ak},
       N_k – number of cycles determined from fatigue characteristics
              $(\sigma_a - N_f)$ or $(\varepsilon_a - N_f)$ for k-th amplitude of stress or strain.

Fatigue life results calculated under the spectral method are marked with T_{SP}. For the case when loading is defined under the matrix of power spectral density of stress, the formula postulated by Bollotin [15] is applicable while for loading defined under the power spectral density of strain the formula (2.24) is applied. The coefficient λ (2.30) from Wirsching [103, 43] is introduced for the modification of computed fatigue life in order to account for the effect of loading frequency spectrum band.

The fatigue life results are presented in Table 5.4 and a graphical interpretation with a bar chart is provided in Fig. 5.18 and 5.19.

Table 5.4: Variance of equivalent quantities and fatigue life calculated by cycle counting and spectral methods

Case of loading	$\mu_{\sigma_{eq}}$, (MPa)^2 and $\mu_{\varepsilon_{eq}}$	$\mu_{G_{\sigma_{eq}}}$, (MPa)^2 and $\mu_{G_{\varepsilon_{eq}}}$	T_{ZC}, s	T_{ZC}^{IFFT}, s	T_{SP}, s
1	2	3	4	5	6
S1	4356	4297	24858958	38013155	22986009
E1	$4,84 \cdot 10^{-8}$	$1,042 \cdot 10^{-7}$	67234162	52705192	52264751
S2	4356	4396	38594548	36551919	43735392
E2	$4,84 \cdot 10^{-8}$	$9,825 \cdot 10^{-8}$	99235609	94869021	124754163
S3	4356	4017	17723845	48991741	29353764
E3	$5,224 \cdot 10^{-8}$	$1,098 \cdot 10^{-7}$	47789773	44796705	46533835
S4	4356	4278	32428800	35294916	47584615
E4	$5,224 \cdot 10^{-8}$	$1,065 \cdot 10^{-7}$	80026925	77494113	101684312
S5	3600	3717	26934505	40972965	38730016
E5	$8,393 \cdot 10^{-8}$	$1,713 \cdot 10^{-7}$	20480584	18811746	17565585
S6	3600	3719	73905786	65152148	80267796
E6	$8,864 \cdot 10^{-8}$	$1,775 \cdot 10^{-7}$	25498783	24941948	33775807
S7	3600	3946	53898992	41783054	31213463
E7	$1,38 \cdot 10^{-7}$	$2,736 \cdot 10^{-7}$	6728184	7543379	6775990
S8	3600	3588	82609952	82569362	90363553
E8	$1,38 \cdot 10^{-7}$	$2,73 \cdot 10^{-7}$	10390489	10394199	14080962
S9	5245	4870	8956385	20632502	14660695
E9	$1,315 \cdot 10^{-7}$	$2,985 \cdot 10^{-7}$	7149073	5590522	5722448
S10	5245	5362	18064104	18824753	21449129
E10	$1,315 \cdot 10^{-7}$	$2,545 \cdot 10^{-7}$	11248256	12324201	16220851

Table 5.4: *(continues)*

1	2	3	4	5	6
S11	3600	3362	50156704	80516261	55567416
E11	$8,163{\cdot}10^{-8}$	$1,912{\cdot}10^{-7}$	19687579	15037548	13995446
S12	3600	3632	81414440	77522673	86324041
E12	$8,165{\cdot}10^{-8}$	$1,642{\cdot}10^{-7}$	30868660	29893642	39199964
S13	3969	3758	47994169	57803042	37250262
E13	$7,29{\cdot}10^{-8}$	$1,618{\cdot}10^{-7}$	27614976	18486880	19815017
S14	3969	4044	52856090	49816318	58751088
E14	$7,29{\cdot}10^{-8}$	$1,45{\cdot}10^{-7}$	38580171	38135528	51519241
S15	4526	4500	26571009	32924271	19509929
E15	$7,502{\cdot}10^{-8}$	$1,423{\cdot}10^{-7}$	24825354	26817960	26163490
S16	4588	4532	32338550	29428628	38878267
E16	$7,631{\cdot}10^{-8}$	$1,551{\cdot}10^{-7}$	34370143	33930882	44757172
S17	5184	5183	16691575	18471462	11692869
E17	$1,175{\cdot}10^{-7}$	$2,233{\cdot}10^{-7}$	9956680	11011236	10157073
S18	5184	5185	17247805	21997937	24257814
E18	$1,175{\cdot}10^{-7}$	$2,399{\cdot}10^{-7}$	13960217	13644716	18154371
S19	5037	5455	12468718	13051272	9755663
E19	$1,667{\cdot}10^{-7}$	$3,324{\cdot}10^{-7}$	4644424	5223805	4640663
S20	4931	4947	22656140	28141801	28511227
E20	$1,552{\cdot}10^{-7}$	$3,104{\cdot}10^{-7}$	8308393	7909212	10941757
S21	4025	3943	37229857	49623582	31272921
E21	$7,673{\cdot}10^{-8}$	$1,546{\cdot}10^{-7}$	23095024	21397256	21837196
S22	4025	4123	49061226	46406646	55077998
E22	$7,673{\cdot}10^{-8}$	$1,545{\cdot}10^{-7}$	35015209	35168103	44793658
S23	3645	3431	37503145	88615708	51673197

Table 5.4: *(continues)*

1	2	3	4	5	6
E23	$9{,}62{\cdot}10^{-8}$	$1{,}861{\cdot}10^{-7}$	14529498	14788695	14744152
S24	3602	3628	74645472	69489813	87332584
E24	$9{,}229{\cdot}10^{-8}$	$1{,}866{\cdot}10^{-7}$	23189441	22747671	30278142

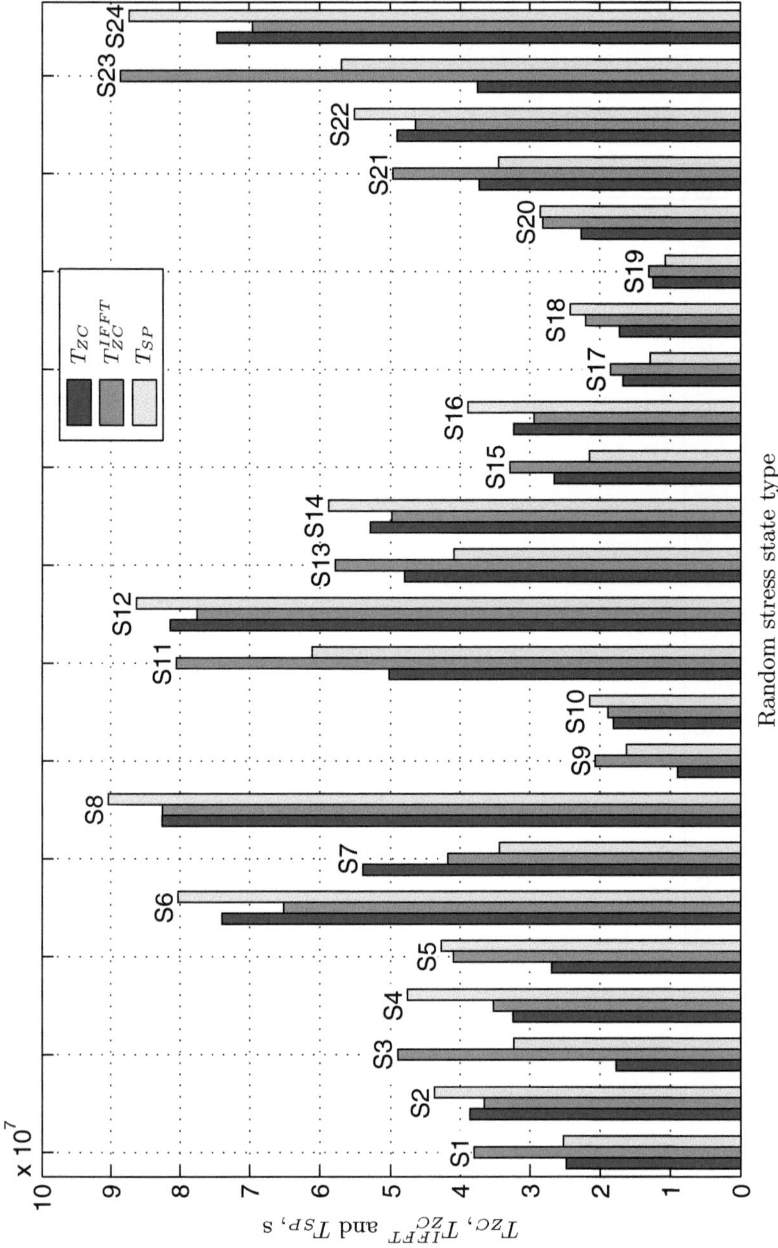

Fig. 5.18. Comparison of fatigue life calculated by cycle counting T_{ZC} and T_{ZC}^{IFFT} and spectral methods T_{SP}

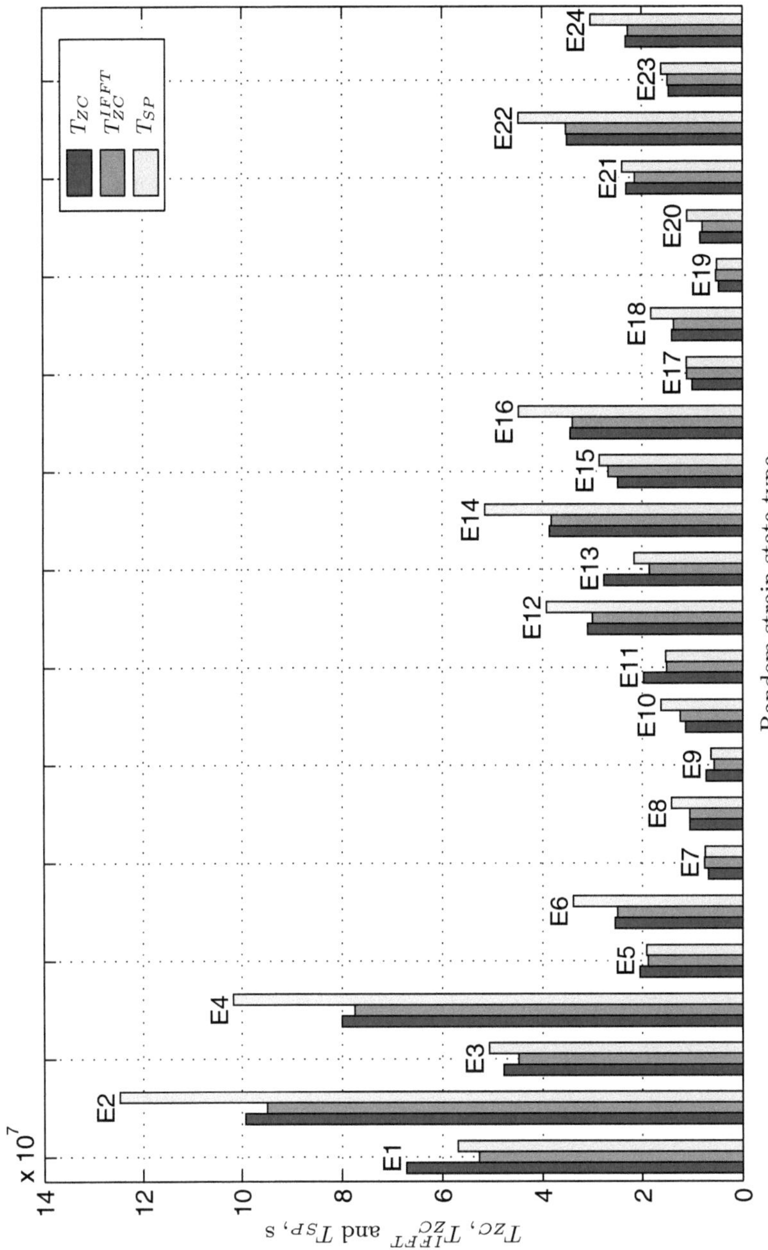

Fig. 5.19. Comparison of fatigue life calculated by cycle counting T_{ZC} and T_{ZC}^{IFFT} and spectral methods T_{SP}

Another technique of fatigue life comparison is presented in Fig. 5.20 $\div$ 5.23. It is remarked that life time T_{ZC}^{IFFT} calculated on the basis of power spectral density of the equivalent stress is overestimated in comparison to life time T_{ZC} for narrow-band frequency spectrum histories (Fig. 5.20). For broad-band frequency spectrum histories, the life time T_{ZC}^{IFFT} fluctuate around T_{ZC} with maximum deviations up to $\pm 9\%$. It could be remarked that it is associated with the generation of random histories by the inverse fast Fourier transform from power spectral density. It results from random determination of phase shifts of the successive harmonic components. Therefore, variations of random histories are registered depending on the number of component harmonics with considerable amplitudes and their phase shifts. The applied generation method is particularly sensitive for the cases of broad-band frequency spectrum of generated histories and correlated tensor components. Additionally, the scatter of the results is affected by the large difference of number of component cycles in narrow- and broad-band frequency spectrums.

The life time calculated in accordance with the generalized spectral method correlates well with life time calculated by the cycle counting method. It should be remarked that the application of various spectral formulae for fatigue life depends on the type and character of the loading (two fatigue characteristics and coefficients λ).

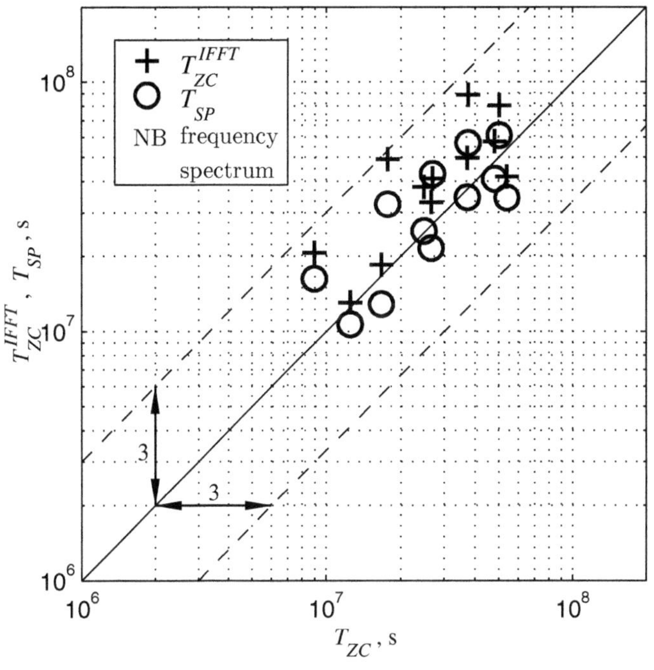

Fig. 5.20.
Comparison
of fatigue life
T_{ZC}^{IFFT} and
T_{SP} with life
calculated by
cycle counting
method T_{ZC}.
Cases of
loading defined
by stress
tensors with
narrow-band
(NB) frequency
spectrum

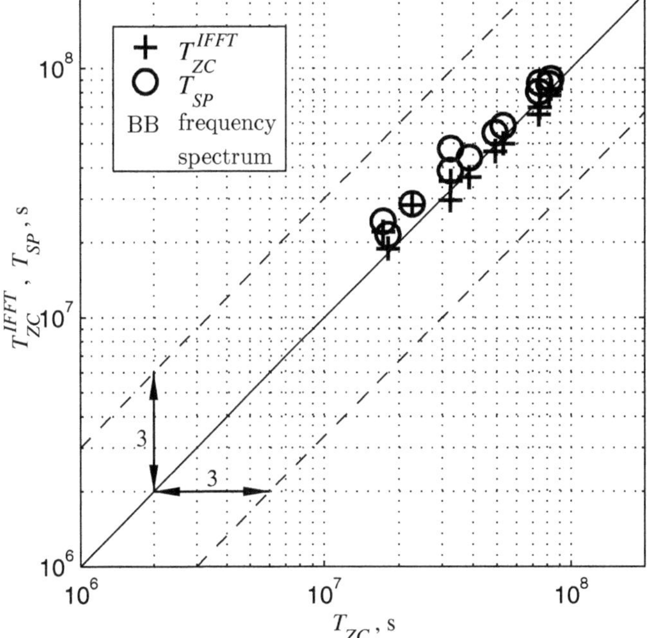

Fig. 5.21.
Comparison
of fatigue life
T_{ZC}^{IFFT} and
T_{SP} with life
calculated by
cycle counting
method T_{ZC}.
Cases of
loading defined
by stress
tensors with
broad-band
(BB) frequency
spectrum

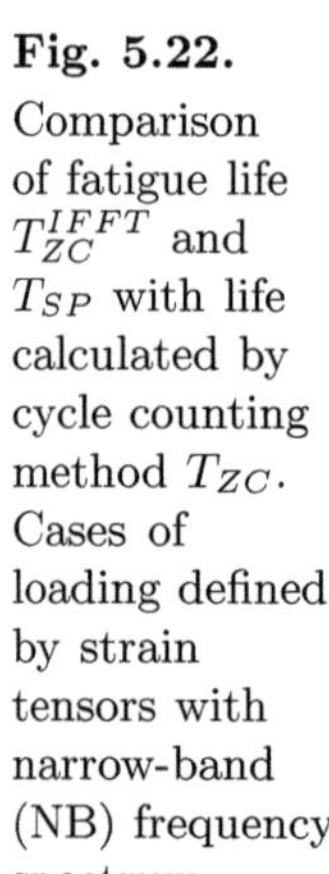

Fig. 5.22. Comparison of fatigue life T_{ZC}^{IFFT} and T_{SP} with life calculated by cycle counting method T_{ZC}. Cases of loading defined by strain tensors with narrow-band (NB) frequency spectrum

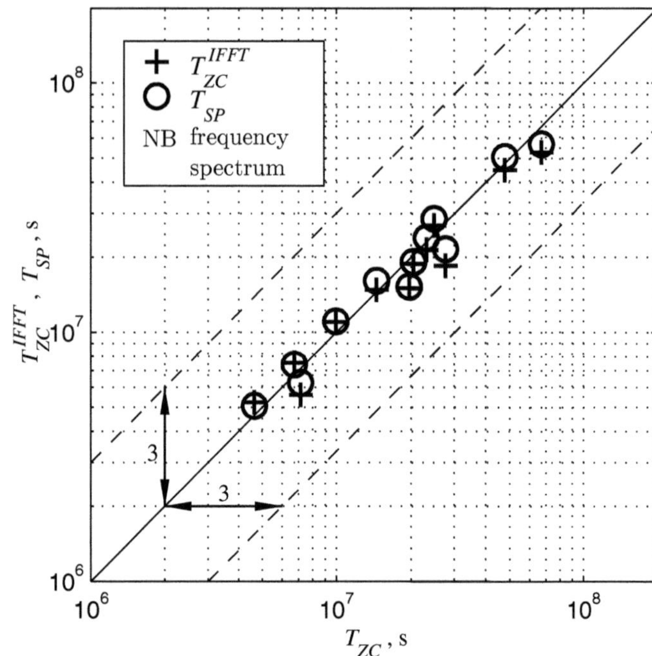

Fig. 5.23. Comparison of fatigue life T_{ZC}^{IFFT} and T_{SP} with life calculated by cycle counting T_{ZC}. Cases of loading defined by strain tensors with broad-band (BB) frequency spectrum

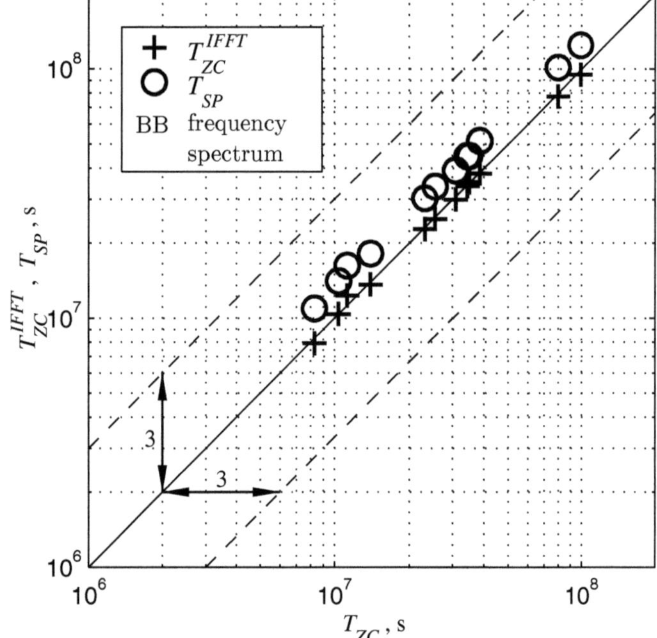

6

Experimental Studies

6.1 Test Stands

Fatigue testing was performed on two stands under the combination of bending with torsion MZGS100L and MZGS200L as presented in Figs 6.1 and 6.2.

The stand MZGS100L is a device designed for fatigue testing of standard specimens made of various materials. The major components include a frame with a rotary table, a lever and an electromagnetic actuator. Under the electrical current the vibrations of the actuator coil generates moment of force M in the tested specimen over the arm. The specimen is fixed at one end to the lever and at the other side is attached to the holder placed at a rotary table (Fig. 6.3 a). Through a suitable rotation over the angle α_M in the limit of $\langle 0 \div \pi/2 \rangle$ the moment M of the tested specimen could be resolved into two component moments-bending M_σ and torsional M_τ. Two extreme cases could be distinguished: when the specimen axis is parallel to the axis of the lever-in which case the specimen loading is limited to bending moment $M = M_\sigma$ and the case in which the axes of the specimen and lever form a right angle and a distinct case of the torsion is involved $M = M_\tau$. Therefore, for the case of MZGS100L stand the tests could involve an arbitrary combination of bending and torsional moments including correlation coefficient $r_{M_\sigma,M_\tau} = 1$, i.e. proportional loading.

Apart from that, the test stand includes a personal computer with input/output card and power amplifier for power supply to electromagnetic actuator. The diagram of the entire stand is presented in Fig. 6.4. The computer-controlled time variable command signal with the voltage range of ± 1 V is generated. The signal is conveyed to the amplifier; and subsequently, the amplified signal is fed into the electromagnetic actuator. The stand includes a limit switch for the control of maximum deflection of the arm. As the preset limit is exceeded the amplifier trips and test timer switches off. As the occurrence of fatigue failure induces a rapid and large loss of specimen rigidity (inclination of the lever increases), a limit switch is applied for the determination of experimental fatigue life T_{exp}.

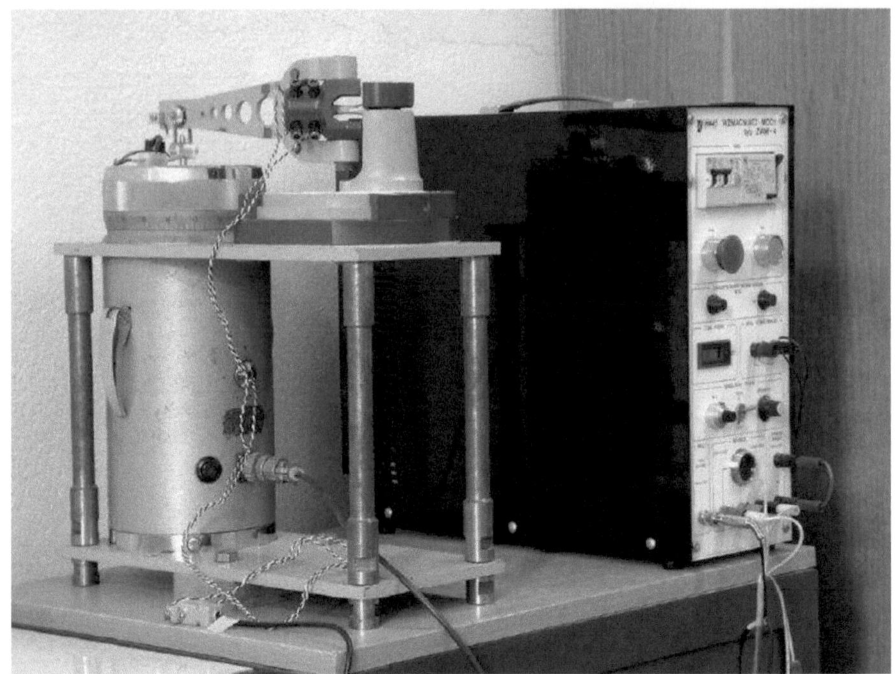

Fig. 6.1. Test stand MZGS100L (PC uplink from below)

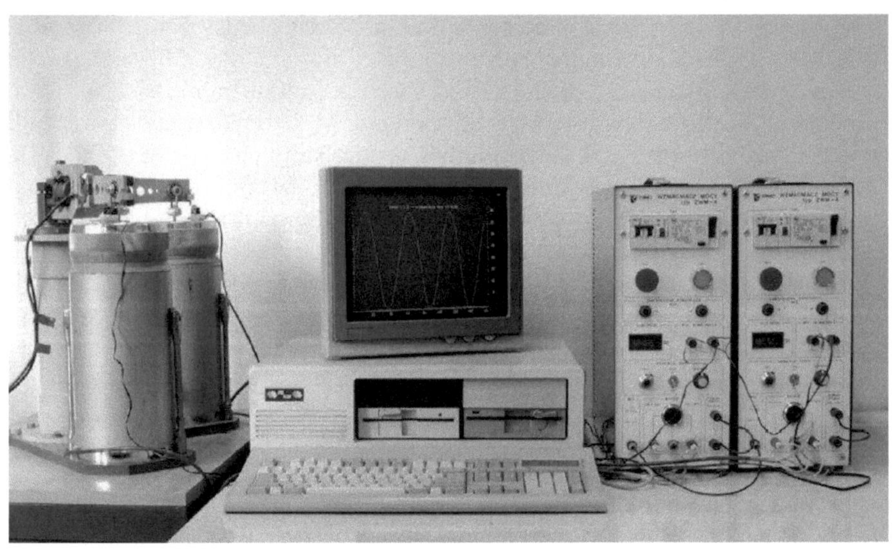

Fig. 6.2. Test stand MZGS200L

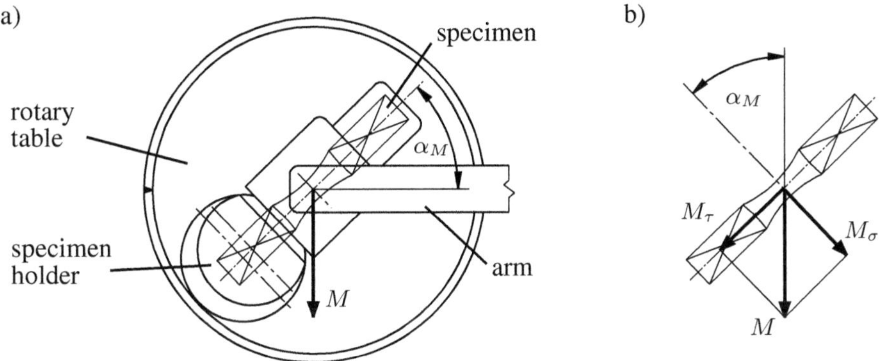

Fig. 6.3. Rotary table of MZGS100L stand with fixed specimen (a) and graphical representation of moment of force M resolved into bending moment M_σ and torsional moment M_τ of specimen (b)

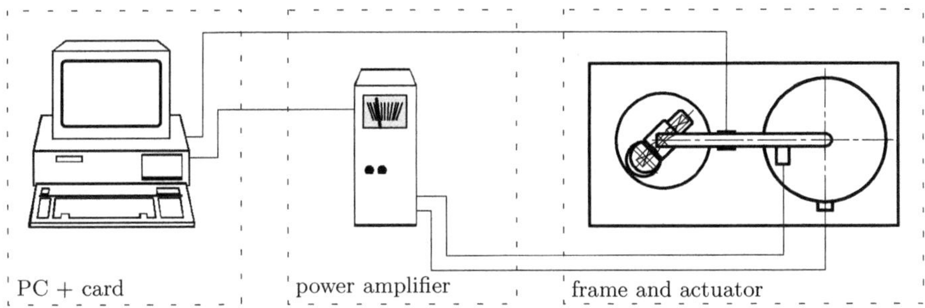

Fig. 6.4. Diagram of connections of components of MZGS100L stand for fatigue testing of specimens under proportional bending and torsion

The measurement of strain at the lever with a strain gauge indicates the instantaneous values of moment M. The voltage at strain bridge of measuring card is scaled on the basis of equation of calibration line. The calibration is performed by the static method by application of weight one after another to the lever and recording the respective voltage. On the basis of the length of the lever the relation between voltage and moment M is determined.

The test stand MZGS200L is a device designed for fatigue testing of material specimens under the non-proportional bending and torsional moments with arbitrary and independent random histories $r_{M_\sigma,M_\tau} = \langle -1 \div 1 \rangle$. In contrast to the MZGS100L stand it is equipped with two independently controlled and supplied electromagnetic actuators. The specimen is fixed to the table and there is no possibility of rotation (Fig. 6.5a). The bending moment M_σ and torsional moment M_τ are equal to the moments from a parallel and perpen-

dicular lever to the axis of the specimen (Fig. 6.5b). The diagram of the stand
is presented in Fig. 6.6.

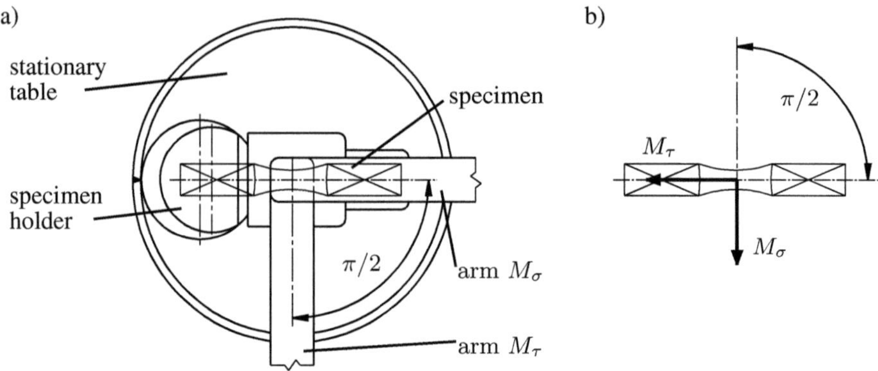

Fig. 6.5. Specimen fixing on table of MZGS200L stand with the angle $\pi/2$ between
arms (a) and graphical interpretation of bending moment M_σ and torsional moment
M_τ of specimen (b)

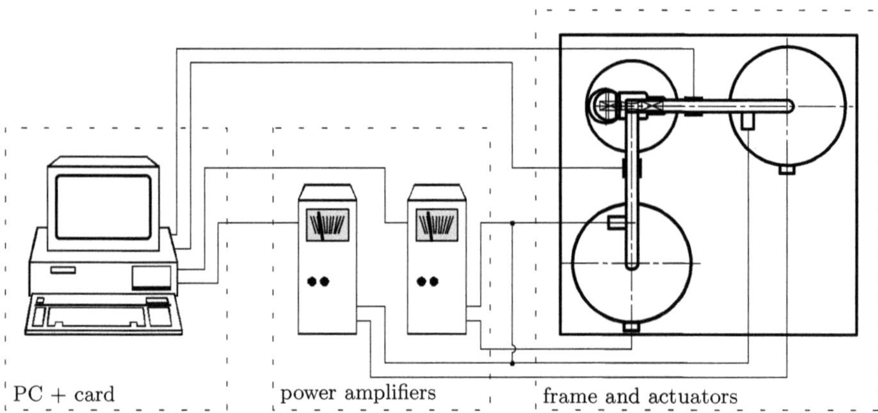

Fig. 6.6. Diagram of connections of components of MZGS200L stand for fatigue
testing of specimens under non-proportional bending and torsion

The operation and control of the MZGS200L stand is similar as in the
case of MZGS100L. The differences result from the necessity of control by
means of two electromagnetic actuators and the measurement of the signals
from two arms simultaneously. Each of the actuators is attached with a limit
switch. The connection of the switches ensures both amplifiers to switch off

in the case when one of the arms exceeds the maximum level of inclination. The bending and torsional arms are statically calibrated and the obtained characteristics served for the determination of adequate moments.

6.2 Material and Geometry of Tested Specimens

Fatigue testing is performed on smooth round specimens made of steel 18G2A. A ground bar is selected as the representative input material for testing. As a result of the assumption that fatigue testing involves marketed structural material (in particular with reference to cast quality) there was not preliminary material selection process. Chemical composition is presented in Table 6.1. The quality of the material complies with the Polish standard of steel products PN-86/H-84018.

Table 6.1. Chemical composition of steel 18G2A (%)

C	Mn	Si	P	S	Cr	Ni	Cu	Fe
0.21	1.46	0.42	0.019	0.046	0.09	0.04	0.17	rest

The specimens are made in accordance with the current standard PN-76/H-04326. Grinding ensures an adequate quality of the mid-surface. The minimum specimen diameter is established with reference to particular tests in the range 6.4-7.5 mm. The shape and dimensions of the specimens are presented in Fig. 6.7. The essential fatigue and strength characteristics of steel 18G2A are presented in Table 5.1 in the chapter 5 devoted to computer simulation.

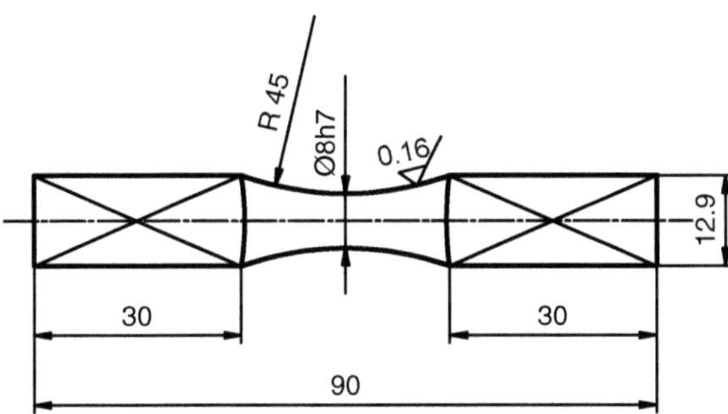

Fig. 6.7. Shape and dimensions of specimens in fatigue test

6.3 Testing under Narrow-Band Frequency Loading

Fatigue investigation under narrow-band Gaussian loading is performed on test stands MZGS100L and MZGS200L. The testing uses a generator of narrow-band frequency history applied during computer simulation. A section of history and the fundamental characteristics are summarized in Fig. 5.2 in the chapter 5. Testing is performed under various value of correlation coefficients $r_{\sigma,\tau}$ between the history of normal stress from bending $\sigma(t)$ and shear stress from torsion $\tau(t)$ and various variance of μ_σ and μ_τ. The results are presented in Table 6.2.

Histories of normal stress $\sigma(t)$ and shear stress $\tau(t)$ are nominal quantities derived from formulae:

$$\sigma(t) = \frac{M_\sigma(t)}{W_x}, \tag{6.1}$$

$$\tau(t) = \frac{M_\tau(t)}{W_0}, \tag{6.2}$$

where: $W_x = \dfrac{\pi d^3}{32}$ – section modulus of bending,

$W_0 = \dfrac{\pi d^3}{16}$ – section modulus of torsion,

d – specimen diameter.

For the case of stand MZGS200L the histories of bending moment $M_\sigma(t)$ and torsional moment $M_\tau(t)$ are direct measure on the appropriate levers. For the case of stand MZGS100L the measurement involves the resulting moment $M(t)$. On the basis of the resulting moment the determination of histories of normal and shear stresses are possible with input of angle α_M:

$$\sigma(t) = \frac{M(t)\cos(\alpha_M)}{W_x}, \tag{6.3}$$

$$\tau(t) = \frac{M(t)\sin(\alpha_M)}{W_0}. \tag{6.4}$$

The fatigue testing under bending, torsion and proportional bending and torsion with correlation coefficient of components $r_{\sigma,\tau} = 1$ are performed entirely on test stand MZGS100L. The selection of the stand provides with a guarantee of constant correlation coefficient and the simple structure ensures reliability and controllability. The stand MZGS200L serves for performing fatigue testing under nonproportional bending and torsion with correlation coefficient $r_{\sigma,\tau} \neq 1$. For the purposes a two-channel generator of random histories with controlled correlation coefficient is prepared. The testing takes advantage of the fact that two various histories $x_1(t)$ and $x_2(t)$ on generator output exhibit lack of correlation, i.e. $r_{x_1,x_2} \approx 0$. By summation of the histories under an appropriate relation a resulting one $x_3(t)$ is obtained in accordance with the relation

$$x_3(t) = x_1(t) + a_r x_2(t) \,, \tag{6.5}$$

where: $x_1(t)$ and $x_2(t)$ – non-correlated histories $r_{x_1,x_2} \approx 0$,

$\quad a_r \qquad\qquad\quad$ – coefficient for determination of correlation.

The following steps involve the selection of a_r coefficient in a way that ensures the required correlation of histories $x_1(t)$ and $x_3(t)$, which are scaled for the same variance

$$x_3(t) = \sqrt{\frac{\mu_{x_1}}{\mu_{x_3}}}\,[x_1(t) + a_r x_2(t)] \,. \tag{6.6}$$

Eight fatigue tests were performed and marked with symbols as follows:

N01 – fatigue testing under pure bending, MZGS100L stand, angle between axis of specimen and lever $\alpha_M = 0°$, 11 specimens,

N02 – fatigue testing under pure torsion, MZGS100L stand, angle between axis of specimen and lever $\alpha_M = 90°$, 5 specimens,

N03 – fatigue testing under nonproportional bending with torsion, MZGS200L stand, correlation coefficient $r_{\sigma,\tau} \approx 0$, relation of standard deviation $\sqrt{\mu_\sigma}/\sqrt{\mu_\tau} \approx 2$, 6 specimens,

N04 – fatigue testing under nonproportional bending with torsion, MZGS200L stand, correlation coefficient $r_{\sigma,\tau} \approx 0$, relation of standard deviation $\sqrt{\mu_\sigma}/\sqrt{\mu_\tau} \approx 1$, 5 specimens,

N05 – fatigue testing under proportional bending with torsion, MZGS100L stand, angle between axis of specimen and lever $\alpha_M = 45°$, correlation coefficient $r_{\sigma,\tau} = 1$, relation of standard deviation $\sqrt{\mu_\sigma}/\sqrt{\mu_\tau} \approx 2$, 8 specimens,

N06 – fatigue testing under proportional bending with torsion, MZGS100L stand, angle between axis of specimen and lever $\alpha_M = 63.5°$, correlation coefficient $r_{\sigma,\tau} = 1$, relation of standard deviation $\sqrt{\mu_\sigma}/\sqrt{\mu_\tau} \approx 1$, 5 specimens,

N07 – fatigue testing under nonproportional bending with torsion, MZGS200L stand, correlation coefficient $r_{\sigma,\tau} \approx 0.5$, relation of standard deviation $\sqrt{\mu_\sigma}/\sqrt{\mu_\tau} \approx 1$, 4 specimens,

N08 – fatigue testing under nonproportional bending with torsion, MZGS200L stand, correlation coefficient $r_{\sigma,\tau} \approx 0.5$, relation of standard deviation $\sqrt{\mu_\sigma}/\sqrt{\mu_\tau} \approx 2$, 5 specimens.

The details of standard deviation of histories $\sigma(t)$ and $\tau(t)$ and results of fatigue life are summarised in Table 6.2.

6.4 Testing under Broad-Band Frequency Loading

Testing under wide band frequency spectrum Gaussian loading was performed on MZGS100L by Achtelik [47]. A section of the registered history along with

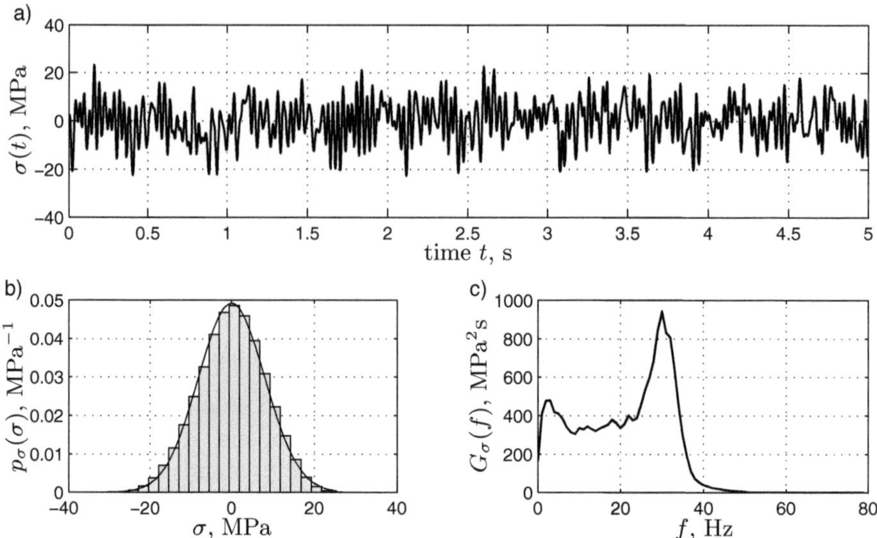

Fig. 6.8. A section of stress history with broad-band frequency (a) with its probability density distribution function (b) and power spectral density function (c)

the probability density function and power spectral density are presented in Fig. 6.8.

The fatigue testing is divided into three groups marked with symbols as follows:

A01 – fatigue testing under pure bending, angle between axis of specimen and lever $\alpha_M = 0°$, 16 specimens,

A02 – fatigue testing under proportional bending with torsion, angle between axis of specimen and lever $\alpha_M = 45°$, relation of standard deviation $\sqrt{\mu_\sigma}/\sqrt{\mu_\tau} \approx 2$, 16 specimens,

A03 – fatigue testing under pure torsion, angle between axis of specimen and lever $\alpha_M = 90°$, 16 specimens.

The details of standard deviation of histories $\sigma(t)$ and $\tau(t)$ and results of fatigue life are summarized in Table 6.2.

6.5 Comparison of Calculated and Experimental Fatigue Life

The fatigue life is computed with the cycle counting and spectral methods. The calculation is performed similarly as in the simulations with the exception of blocks PWO1, PWO2, PWO3 applied for comparison. The diagram of the algorithm of life time computation is presented Fig. 6.9.

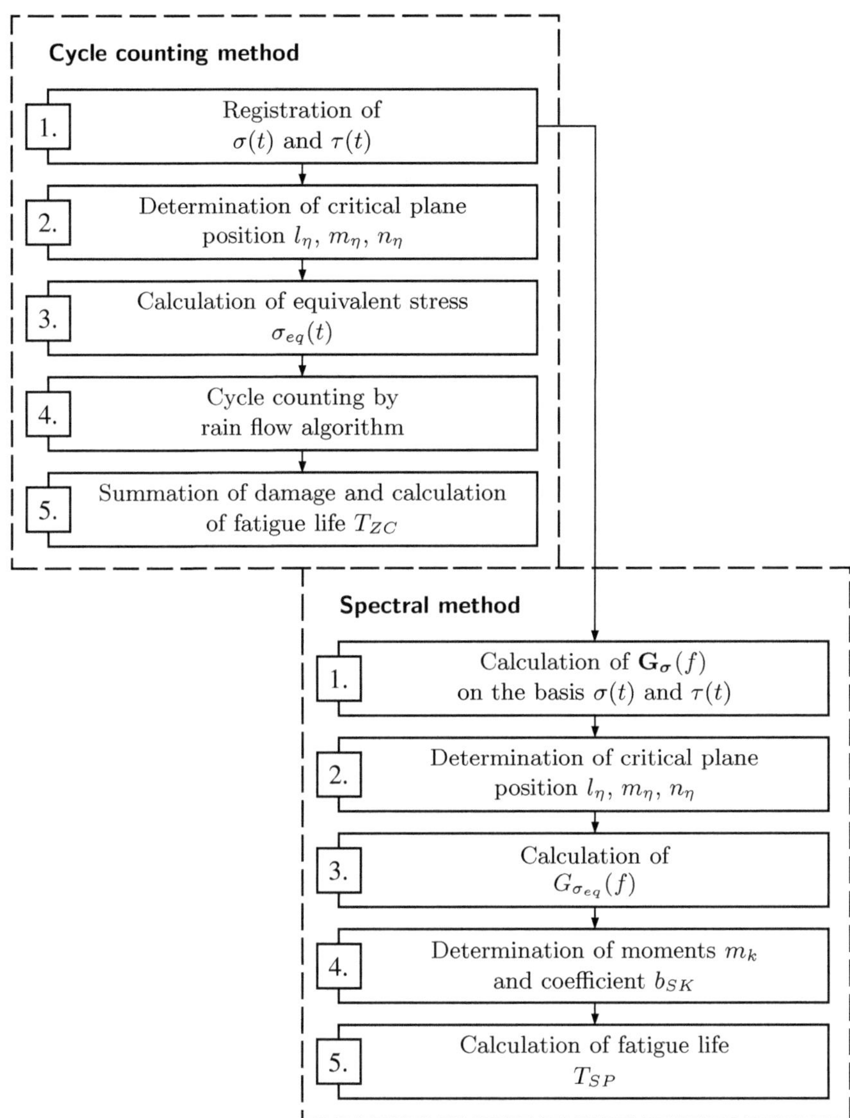

Fig. 6.9. Algorithm of fatigue life computation by cycle counting and spectral methods

In order to simplify further calculations it is assumed that the consider-ations involve plain state of stress for which case the vector of tensor com-ponents takes the form $\boldsymbol{\sigma}(t) = [\sigma(t), 0, \tau(t)]$. Random histories of nominal stresses resulting from bending $\sigma(t)$ and torsion $\tau(t)$ determined from (6.1) and (6.3) constitute the basis for the calculations. The modified criterion of maximum normal stress in the critical plane [48] is applied for the determi-nation of the history of equivalent stress, which takes the form

$$\sigma_{eq}(t) = l_\eta^2 \sigma(t) + 2l_\eta m_\eta \frac{\sigma_{af}}{\tau_{af}} \tau(t) . \tag{6.7}$$

The modification involves the consideration of the difference of levels of fatigue limits of bending σ_{af} and torsion τ_{af}. For the plain state of stress the position of the critical plane could be defined as a function of a single direction cosines, as for $n_\eta = 0$ we obtain $m_\eta = \sqrt{1 - l_\eta^2}$. It is a remark, which assists in the calculations and reduces its duration. Table 6.2 presents the value of direction cosines l_η of vector normal to the critical plane defined by the maximum variance method. The determination of critical plane in time and frequency domains leads to identical results.

The equivalent stress history established from (6.7) is submitted to cycle counting process with the rain flow algorithm. The derived cycles serve for damage accumulation by the linear Serensen-Kogayev hypothesis (3.79) [51, 88, 95]. The fatigue life formula is derived in the form

$$T_{ZC} = \frac{T_o b_{SK} A}{\sum\limits_{i=1}^{k} n_i \sigma_{ai}^m} \qquad \text{for} \quad \sigma_{ai} \geq a_{SK} \sigma_{af} . \tag{6.8}$$

Up to the present the Serensen-Kogayev hypothesis of damage accumu-lation has not been applied in the spectral method for correlation of exper-imental data under random loading with broad-band frequency spectrum. Therefore, it is necessary to derive a formula for fatigue life T_{SP} under the assumption that loading history has normal distribution with broad-band fre-quency spectrum. The general notation for this case is following

$$T_{SP} = \frac{1}{\lambda} T_{NB} , \tag{6.9}$$

where: λ – coefficient for accounting the effect of width of band frequency of loading on fatigue life [103],

T_{NB} – fatigue life under the assumption of narrow-band frequency spectrum and Serensen-Kogayev damage accumulation hypoth-esis (3.82).

However, the fact that b_{SK} coefficient depends on the variance of the equivalent history μ_σ and expected maximum of cycle amplitude $\sigma_{a\,max}$ must be considered. The quantities are different for each of the performed tests; therefore, b_{SK} coefficient is determined separately for each case. The difficulty consists in the determination of maximum of cycle amplitude $\sigma_{a\,max}$ directly from power spectral density of equivalent stress history. In practice, the maximum amplitude of a cycle has a limited value. In the investigated case, following a number of laboratory tests and statistical processing the following formula is derived

$$\sigma_{a\,max} = 3.73\sqrt{\mu_\sigma}\,.\tag{6.10}$$

The results of fatigue life calculations by the cycle counting method T_{ZC} (6.8) and spectral method T_{SP} (6.9) are summarised in Table 6.2.

Table 6.2: Results of fatigue testing under combination of bending with torsion and calculated fatigue life

Symbol and number		I	$r_{\sigma,\tau}$	$\sqrt{\mu_\sigma}$ MPa	$\sqrt{\mu_\tau}$ MPa	$\sqrt{\mu_\sigma/\mu_\tau}$	T_{exp} s	T_{ZC} s	T_{SP} s	l_η
1		2	3	4	5	6	7	8	9	10
N01	1	0.99	–	150.6	–	–	13560	16605	14777	0
	2	0.99	–	127.3	–	–	55500	54882	48154	0
	3	0.99	–	149.4	–	–	9660	17520	15583	0
	4	0.99	–	113.4	–	–	195720	125910	109170	0
	5	0.99	–	133.2	–	–	43260	39591	34879	0
	6	0.99	–	134.4	–	–	43080	37275	32857	0
	7	0.99	–	134.4	–	–	35340	37275	32857	0
	8	0.99	–	139.8	–	–	21240	28140	24911	0
	9	0.99	–	111.1	–	–	316200	145560	126042	0
	10	0.99	–	133.0	–	–	33660	40215	35407	0
	11	0.99	–	128.1	–	–	62100	52367	45942	0
N02	1	0.99	–	–	66.0	–	448020	267681	230396	0.707
	2	0.99	–	–	63.0	–	653760	369358	316066	0.707
	3	0.99	–	–	63.0	–	226800	369358	316066	0.707
	4	0.99	–	–	64.0	–	352200	324421	277995	0.707
	5	0.99	–	–	65.0	–	219360	285198	245131	0.707
N03	1	0.99	-0.01	96.8	54.2	1.79	191880	193105	194656	0.883
	2	0.99	-0.01	96.8	54.2	1.79	144240	193105	194656	0.883
	3	0.99	-0.01	118.3	66.2	1.79	30600	46724	46797	0.883
	4	0.99	-0.01	118.3	66.3	1.78	35100	46724	46797	0.883
	5	0.99	-0.01	106.5	59.6	1.79	118980	98309	99046	0.883

Table 6.2: *(continuation)*

		1	2	3	4	5	6	7	8	9	10
	6	0.99	-0.01	104.0	58.2	1.79	141300	116763	117660	0.883
N04	1	0.99	-0.01	57.9	56.2	1.03	223320	470001	435709	0.766
	2	0.99	-0.01	57.9	56.2	1.03	416400	470001	435709	0.766
	3	0.99	-0.01	57.9	56.2	1.03	295740	470001	435709	0.766
	4	0.99	-0.01	63.7	61.8	1.03	168360	237306	219352	0.766
	5	0.99	-0.01	57.9	56.2	1.03	303660	470001	435709	0.766
N05	1	0.99	1	108.8	54.4	2	20700	13739	11906	0.866
	2	0.99	1	108.0	54.0	2	22800	14517	12581	0.866
	3	0.99	1	107.4	53.7	2	20580	15073	13056	0.866
	4	0.99	1	96.3	48.1	2	41820	32628	28040	0.866
	5	0.99	1	100.0	50.0	2	25020	24965	21526	0.866
	6	0.99	1	89.2	44.6	2	81720	56314	48072	0.866
	7	0.99	1	92.0	46.0	2	92280	45113	38549	0.866
	8	0.99	1	79.8	39.9	2	380220	124908	105339	0.866
N06	1	0.99	1	61.8	61.8	1	134700	41552	35563	0.788
	2	0.99	1	61.8	61.8	1	155520	41552	35563	0.788
	3	0.99	1	61.8	61.8	1	99600	41552	35563	0.788
	4	0.99	1	61.8	61.8	1	148740	41552	35563	0.788
	5	0.99	1	61.8	61.8	1	167040	41552	35563	0.788
N07	1	0.99	0.5	71.7	69.6	1.03	48540	32955	30944	0.766
	2	0.99	0.5	64.3	62.4	1.03	205140	70376	66190	0.766
	3	0.99	0.5	63.4	61.5	1.03	72600	77879	73365	0.766

Table 6.2: *(continuation)*

	1	2	3	4	5	6	7	8	9	10
	4	0.99	0.5	63.4	61.5	1.03	119700	77879	73365	0.766
N08	1	0.99	0.5	118.5	64.0	1.85	20940	12817	13073	0.848
	2	0.99	0.5	100.7	54.4	1.85	86820	37959	38935	0.848
	3	0.99	0.5	92.3	49.8	1.85	399000	69310	71383	0.848
	4	0.99	0.5	101.3	54.7	1.85	133200	35816	36708	0.848
	5	0.99	0.5	100.8	54.4	1.85	78180	37959	38935	0.848
A01	1	0.78	–	240.9	–	–	5700	5312	3335	0
	2	0.78	–	240.9	–	–	4920	5312	3335	0
	3	0.78	–	240.9	–	–	6480	5312	3335	0
	4	0.78	–	189.6	–	–	17880	12856	9803	0
	5	0.78	–	189.6	–	–	17160	12856	9803	0
	6	0.78	–	189.6	–	–	14520	12856	9803	0
	7	0.78	–	142.2	–	–	19860	44097	34716	0
	8	0.78	–	142.2	–	–	19560	44097	34716	0
	9	0.78	–	142.2	–	–	27960	44097	34716	0
	10	0.78	–	93.1	–	–	98520	267684	256509	0
	11	0.78	–	93.1	–	–	114360	267684	256509	0
	12	0.78	–	93.1	–	–	173820	267684	256509	0
	13	0.78	–	93.1	–	–	218940	267684	256509	0
	14	0.78	–	69.1	–	–	330000	1476985	1339508	0
	15	0.78	–	69.1	–	–	330000	1476985	1339508	0
	16	0.78	–	69.1	–	–	330000	1476985	1339508	0
A02	1	0.78	1	170.4	85.2	2	3060	2132	1151	0.875

Table 6.2: (continuation)

1	2	3	4	5	6	7	8	9	10
2	0.78	1	170.4	85.2	2	1920	2132	1151	0.875
3	0.78	1	170.4	85.2	2	3900	2132	1151	0.875
4	0.78	1	134.0	67.0	2	6600	6557	4710	0.875
5	0.78	1	134.0	67.0	2	10560	6557	4710	0.875
6	0.78	1	117.6	58.8	2	13980	14381	10072	0.875
7	0.78	1	117.6	58.8	2	15480	14381	10072	0.875
8	0.78	1	117.6	58.8	2	18900	14381	10072	0.875
9	0.78	1	100.6	50.3	2	39360	31716	24860	0.875
10	0.78	1	100.6	50.3	2	26040	31716	24860	0.875
11	0.78	1	100.6	50.3	2	29880	31716	24860	0.875
12	0.78	1	100.6	50.3	2	22320	31716	24860	0.875
13	0.78	1	83.4	41.7	2	108300	96521	73569	0.875
14	0.78	1	83.4	41.7	2	143700	96521	73569	0.875
15	0.78	1	65.8	32.9	2	330000	309305	291157	0.875
16	0.78	1	65.8	32.9	2	300000	309305	291157	0.875
A03 1	0.78	—	—	97.9	—	5100	21498	9250	1
2	0.78	—	—	97.9	—	6000	21498	9250	1
3	0.78	—	—	86.2	—	10920	43686	22064	1
4	0.78	—	—	86.2	—	15960	43686	22064	1
5	0.78	—	—	86.2	—	18600	43686	22064	1
6	0.78	—	—	77.1	—	43740	83505	51868	1
7	0.78	—	—	77.1	—	35160	83505	51868	1
8	0.78	—	—	77.1	—	41100	83505	51868	1

Table 6.2: *(continuation)*

1	2	3	4	5	6	7	8	9	10
9	0.78	–	–	67.6	–	103200	222159	132326	1
10	0.78	–	–	67.6	–	159000	222159	132326	1
11	0.78	–	–	67.6	–	62400	222159	132326	1
12	0.78	–	–	62.8	–	117300	317010	226201	1
13	0.78	–	–	62.8	–	276180	317010	226201	1
14	0.78	–	–	62.8	–	205320	317010	226201	1
15	0.78	–	–	62.8	–	300000	317010	226201	1
16	0.78	–	–	57.8	–	330000	596652	401620	1

I – irregularity coefficient,

$r_{\sigma,\tau}$ – correlation coefficient,

$\sqrt{\mu_\sigma}$ – standard deviation of stress resulting from bending,

$\sqrt{\mu_\tau}$ – standard deviation of stress resulting from torsion,

T_{exp} – experimental fatigue life,

T_{ZC} and T_{SP} – fatigue life calculated with cycle counting and spectral method, respectively,

l_η – direction cosines of vector normal to critical plane in relation to x axis of specimen.

The comparison of calculated fatigue life T_{ZC} and T_{SP} with experiment results is presented in Figs 6.10–6.16. The quantities on the axes of the figures are presented in a logarithmic order. The dashed lines mark the scatter band with coefficient 3 which indicates that the points within the band result from the comparison of two quantities for which the equation $1/3 < y/x < 3$ is satisfied. The diagonal (full line) defines an ideal case for which the two compared quantities $y/x = 1$.

For the case of narrow-band frequency loading under pure bending (N01), pure torsion (N02) in Fig. 6.10 and nonproportional bending with torsion with correlation coefficient $r_{\sigma,\tau} \approx 0$ (N03 and N04) in Fig. 6.11 the results are contained in the acceptable scatter band. In Figs 6.12 and 6.13 a comparison is made between the experimental and computed results for correlation coefficient $r_{\sigma,\tau} = 1$ (N05 and N06) and $r_{\sigma,\tau} \approx 0.5$ (N07 and N08). For the latter case the fatigue life calculated with the cycle counting method and spectral method is lower with reference to experimental life.

The results of fatigue life for loading with broad-band frequency spectrum are compared with experiments in Fig. 6.14 and 6.16 For this cases the best equivalence of results is obtained for the case of life calculated by the spectral method (A03 – pure torsion).

During fatigue testing under the combination of bending with torsion performed for various ratios of standard deviation $\sqrt{\mu_\sigma/\mu_\tau} \approx 1$ and ≈ 2 the effect of the ratio on the difference of calculated T_{ZC}, T_{SP} and experimental life T_{exp} is not observed.

The fatigue life results T_{ZC} and T_{SP} calculated for narrow-band frequency scpectrum (N01–N08) are similar with the maximum difference up to 16%. For the histories with broad-band frequency spectrum (A01–A03) the maximum differences amount to higher values (up to 43%), which could be associated with the application of misfiled coefficient λ accounting for the frequency width spectrum of stress.

In general, the spectral method tends to underestimation of fatigue life in comparison to the cycle counting method. It is commonly considered as an advantage of the method [93] due to the estimation of shorter fatigue life. Therefore, the term safety error is used.

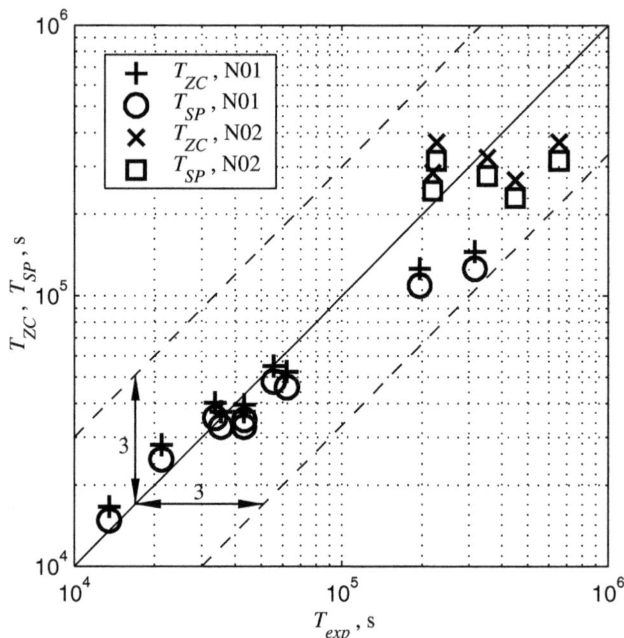

Fig. 6.10.

Comparison of fatigue life calculated by cycle counting T_{ZC} and spectral T_{SP} methods with experimental life T_{exp} under pure bending N01 end pure torsion N02 (narrow-band frequency spectrum)

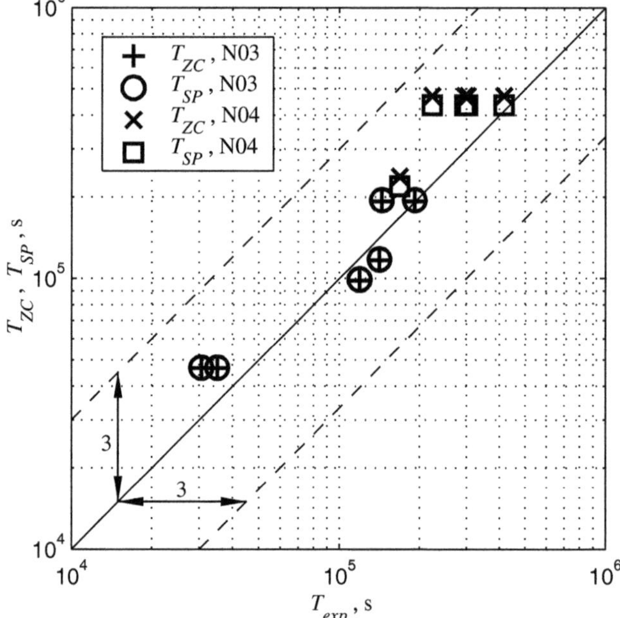

Fig. 6.11.

Comparison of fatigue life calculated by cycle counting T_{ZC} and spectral T_{SP} methods with experimental life T_{exp} under nonproportional bending with torsion N03 and N04 ($r_{\sigma,\tau} \approx 0$) (narrow-band frequency spectrum)

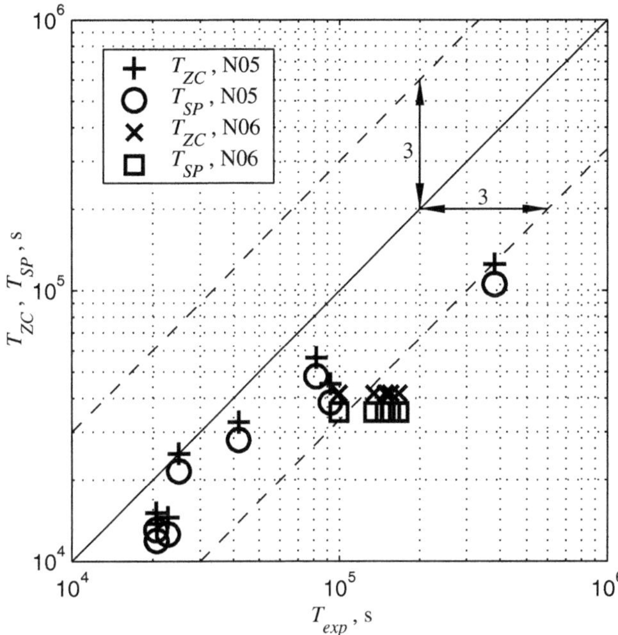

Fig. 6.12. Comparison of fatigue life calculated by cycle counting T_{ZC} and spectral T_{SP} methods with experimental life T_{exp} under proportional bending with torsion N05 and N06 ($r_{\sigma,\tau} = 1$) (narrow-band frequency spectrum)

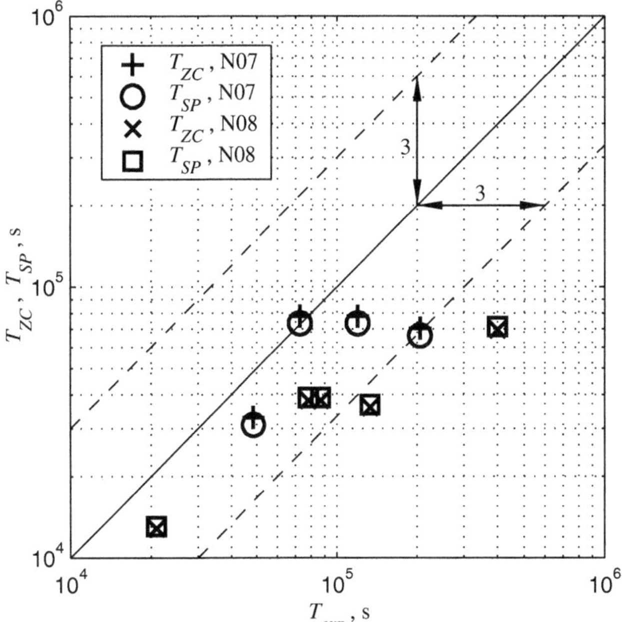

Fig. 6.13. Comparison of fatigue life calculated by cycle counting T_{ZC} and spectral T_{SP} methods with experimental life T_{exp} under nonproportional bending with torsion N07 and N08 ($r_{\sigma,\tau} \approx 0.5$) (narrow-band frequency spectrum)

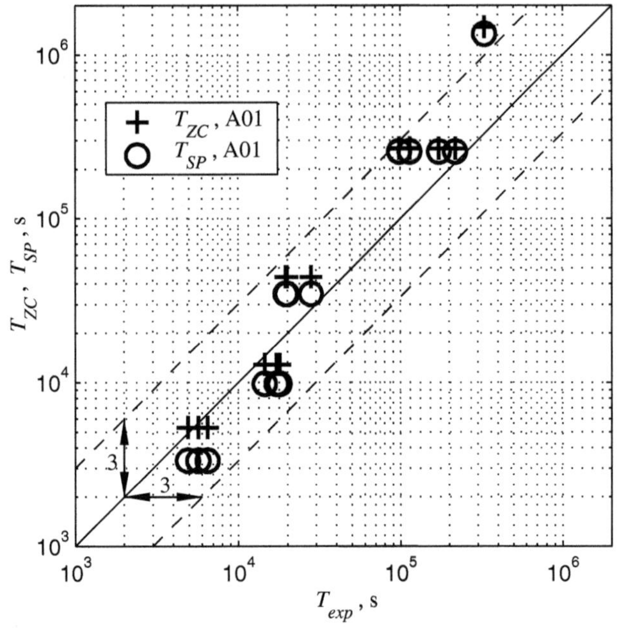

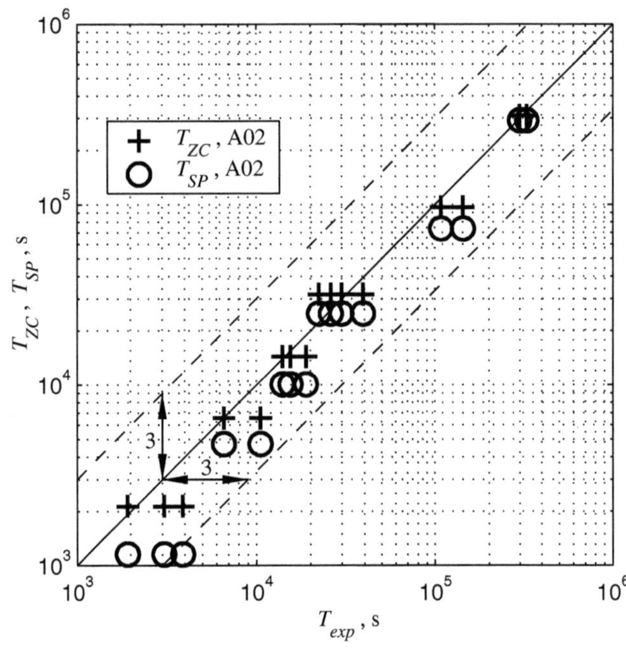

Fig. 6.14.

Comparison of fatigue life calculated by cycle counting T_{ZC} and spectral T_{SP} methods with experimental life T_{exp} under pure bending A01 (broad-band frequency spectrum)

Fig. 6.15.

Comparison of fatigue life calculated by cycle counting T_{ZC} and spectral T_{SP} methods with experimental life T_{exp} under proportional bending with torsion A02 ($r_{\sigma,\tau} = 1$) (broadband frequency spectrum)

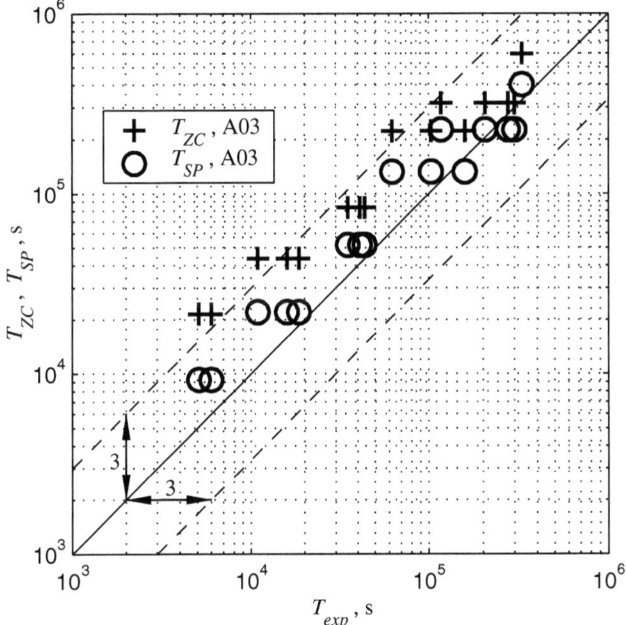

Fig. 6.16. Comparison of fatigue life calculated by cycle counting T_{ZC} and spectral T_{SP} methods with experimental life T_{exp} under pure torsion A03 (broad-band frequency spectrum)

7

Conclusions

1. A generalised spectral method for fatigue life calculation of materials under multiaxial random loading defined in frequency domain expressed in terms of power spectral density functions is postulated.

2. The generalised spectral method is based on linear strain and stress criteria of fatigue failure which enable the reduction of multiaxial stress state to the equivalent uniaxial one.

3. In the equivalent uniaxial stress state fatigue damage is accumulated with use of standard material characteristics determined in stresses for high-cycle fatigue regime and in strains for high- and low-cycle fatigue life regimes.

4. Calculations and analyses of various stimulated random stress and strain states indicate that:

   4.1. The application of linear criteria of multiaxial fatigue failure for the determination of fatigue life in frequency domain by the spectral method gives practically identical results as the cycle counting method in time domain.

   4.2. Equivalent results are gained by the variance method for the determination of the expected critical plane position based on random histories and the power spectral density functions.

5. Fatigue testing of 18G2A steel specimens under proportional and non-proportional random bending with torsion under narrow- and broad-band frequency spectra and zero expected values indicate that:

   5.1. Fatigue life calculated with the generalised spectral method, whose power spectral density of the equivalent stress is carried out by means of the modified criterion of maximum normal stress in the critical plane and damage accumulation is determined in accordance with

Serensen-Kogayev hypothesis, extends into the acceptable scatter band with coefficient 3 determined by experimental life.

5.2. Similar fatigue life is gained from the generalised spectral method and the range counting method. The latter one is more time consuming.

6. The postulated spectral method offers considerable prospects of its applicability in association with the finite element method for mapping life time of structural materials under multiaxial service loading.

References

1. ASTM E 1049–85 (Reapproved 1997): *Standard practices for cycle counting in fatigue analysis*, in: Annual Book of ASTM Standards, Vol. 03.01, Philadelphia 1999, pp. 710–718

2. Banvillet A., Łagoda T., Macha E., Niesłony A., Palin-Luc T., Vittori J.-F.: *Fatigue life under non-Gaussian random loading from various models*, International Journal of Fatigue, Vol. 26, No 4, 2004, pp. 349-363

3. Benasciutti D.: *Fatigue Analysis of Random Loadings*, PhD Thesis, Department of Engineering, University of Ferrara, Italy, 2005, pp. 254

4. Benasciutti D., Tovo R.: *Modelli di previsione del danneggiamento a fatica per veicoli in regime stazionario ed ergodico*, Associazione Italiana per l'Analisi delle Sollecitazioni (AIAS), XXXI Convegno Nazionale – 18- 21 Settembre 2002, Parma, p. 10

5. Benasciutti D., Tovo R.: *Spectral methods for lifetime prediction under wide-band stationary random processes*, Cumulative Fatigue Damage, Seville, 2003, p. 5

6. Bendat J.S., Piersol A.G.: *Random Data. Analysis and Measurement Procedures*, John Wiley and Sons Inc., New York, 1976

7. Bendat J.S., Piersol A.G.: *Engineering Applications of Correlation and Spectral Analysis*, John Wiley and Sons Inc., New York, 1980

8. Będkowski W., Lachowicz C., Macha E.: *Predicted fatigue fracture plane according to variance method of shear stress under random triaxial stress state*, Failure Analysis – Theory and Practice, Proceedings of the 7th European Conference on Fracture (ECF 7), Budapest 1988, Ed. E. Czoboly, EMAS U.K., Vol. I, pp. 281–283

9. Będkowski W., Macha E.: *Fatigue fracture plane under multiaxial random loadings – prediction by variance of equivalent stress based on the maximum shear and normal stress*, Mat. –wiss. u. Werkstofftech., Vol. 23, 1992, pp. 82–94

10. Bílý M., Bukoveczky J.: *Digital simulation of environmental processes with respect to fatigue*, Journal of Sound and Vibration, Vol. 49, No. 4, 1976, pp. 551–568

11. Bishop N.W.M.: *Spectral method of estimating the integrity of structural components subjected to random loading*, Handbook of Fatigue Crack Propagation in Metallic Structures, Ed. An. Carpinteri, Elsevier Science B.V., 1994, pp. 1685–1720

12. Bishop N.W.M., Sherratt F.: *Fatigue life prediction from power spectral density data. Part 2: recent developments*, Environmental Engineering, Vol. 2, 1989, pp. 11–19

13. Bishop N.W.M., Sherratt F.: *A theoretical solution for the estimation of "rainflow" ranges from power spectral density data*, Fatigue and Fracture of Engineering Materials and Structures, Vol. 13, No. 4, 1990, pp. 311–326

14. Bitner- Gregersen E.M., Cramer E.H.: *Uncertainties of load characteristics and fatigue damage of ship structures*, Marine Structures, Vol. 8, 1995, pp. 97–117

15. Bolotin V.V.: *Random Vibration of Elastic Systems*, Martinus Nijhoff Publishers, Hague, The Netherlands, 1984, p. 480

16. Bruder T., Heuler P., Klätschke H., Störzel K.: *Analysis and synthesis of standardized multi-axial load-time histories for structural durability assessment*, Proc. of the Seventh Int. Conf. on Biaxial/Multiaxial Fatigue and Fracture, Berlin, Germany, pp. 63–77

17. Chaudhury G.K., Dover W.D.: *Fatigue analysis of offshore platforms subject to sea wave loadings*, International Journal of Fatigue, Vol. 7, No. 1, 1985, pp. 13–19

18. Chow C.L., Li D.L.: *An analytical solution for fast fatigue assessment under wide-band random loading*, International Journal of Fatigue, 1991, pp. 395–404

19. Clevenson S.A., Stainer R.: *Fatigue Life under Random Loading for Several Power Spectral Shapes*, NASA TR R- 266, Langley Research Center, 1983, p. 18

20. Corten H.T., Dolan T.L.: *Cumulative fatigue damage*, International Conference on Fatigue of Metals, ASME and IME, London, 1956, p. 235–246

21. Dirlik T.: *Application of Computers in Fatigue Analysis*, PhD Thesis, UK: University of Warwick, 1985

22. Dowling N.E.: *Fatigue failure prediction for complicated stress- strain histories*, Journal of Materials, Vol. 7, No. 1, 1972, pp. 71–87

23. Etube L.S., Brennan F.P., Dover W.D.: *Modelling of jack- up response for fatigue under simulated service condition*, Marine Structures, Vol. 12, 1999, pp. 327–348

24. Folsø R.: *Spectral fatigue damage calculation in the side shells of ships, with due account taken of the effect of alternating wet dry areas*, Marine Structures, Vol. 11, 1998, pp. 319–343

25. Friis Hansen P., Winterstein S.R.: *Fatigue damage in the side shells of ships*, Marine Structures, Vol. 8, 1995, pp. 631–655

26. Fu T.-T., Cebon D.: *Predicting fatigue lives for bi- modal stress spectral densities*, International Journal of Fatigue, Vol. 22, 2000, pp. 11–21

27. Gasiak G., Pawliczek R.: *Calculation of fatigue life of specimens made of 18G2A steel under variable bending and torsion loading with different stress ratios,*

Scientific Papers of Technical University of Opole, Vol. 66, No. 268, Opole 2001, pp. 23–44, (in Polish)

28. Grzelak J., Łagoda T., Macha E.: *Spectral analysis of the criteria for multiaxial random fatigue*, Mat. –wiss. u. Werkstofftech., Vol. 22, 1991, pp. 85-98

29. Haibach E.: *Betriebsfestigkeit, Verfaren und Daten zur Bauteilberechnung*, VDI, Düsseldorf, 1989, p. 141

30. Hartt W.H., Lin N.K.: *A proposed stress history for fatigue testing applicable to offshore structures*, International Journal of Fatigue, Vol. 8, No. 2, 1986, pp. 91–93

31. Heuler P., Birk O., Beste A.: *Auf dem Weg zur virtuellen Festigkeit im Fahrzeugbau*, Mat. –wiss. u. Werkstofftech., Vol. 32, 2001, pp. 369–376

32. Holmes J.D: *Fatigue life under along- wind loading — closed- form solutions*, Engineering Structures, Vol. 24, 2002, pp. 109–114

33. Jiao G.: *A theoretical model for the prediction of fatigue under combined Gaussian and impact load*, International Journal of Fatigue, Vol. 17, No. 3, 1995, pp. 215–219

34. Johannesson P.: *Rainflow Analysis of Switching Markov Loads*, Mathematical Statistics, Centre for Mathematical Sciences, Lund Institute of Technology, Sweden, Doctoral Theses in Mathematical Sciences 1999:4, ISBN 91-628-3784-2, p. 195

35. Kam J.C.P., Dover W.D.: *Mathematical background for applying multiple axes random stress histories in the fatigue testing of offshore tubular joints*, International Journal of Fatigue, Vol. 11, No. 5, 1989, pp. 319–326

36. Karolczuk A., Macha E.: *A review of critical plane orientations in multiaxial fatigue failure criteria of metallic materials*, International Journal of Fracture, Vol. 134, 2005, pp. 267–304

37. Korn G.A., Korn T.M.: *Mathematical Handbook*, 2nd ed., McGraw- Hill Book Company, New York, 1968

38. Kowalewski J.: *On the Relationship between Component Life under Irregularly Fluctuating and Ordered Load Sequences, Part 2*, DVL Report 249, 1963

39. Lachowicz C.T., Łagoda T., Macha E.: *Comparison of analytical and algorithmical methods for life time estimation in 10HNAP steel under random loadings*, FATIGUE '96, Proc. of the Sixth Int. Fatigue Congress, (6IFC) Berlin, Eds G. Lütjering and H. Nowack, Pergamon 1996, Vol. 1, pp. 595–600

40. Larsen C.E., Lutes L.D.: *Predicting the fatigue life of offshore structures by the single- moment spectral method*, Probabilistic Engineering Mechanics, Vol. 6, No. 2, 1991, pp. 96–108

41. Leser Ch., Juneja L., Thangjitham S., Dowling N.E.: *On multi- axial random fatigue load modeling*, SAE Technical Paper Series, No. 980696, Reprinted from: Advancements in Fatigue Research and Applications (SP- 1341), 1998, pp. 93–106

42. Liou H.Y., Wu W.F., Shin C.S.: *A modified model for the estimation of fatigue life derived from random vibration theory*, Probabilistic Engineering Mechanics, Vol. 14, 1999, pp. 281–288

43. Liu H.J., Hu S.R.: *Fatigue under nonnormal random stresses using Monte-Carlo method*, FATIGUE' 87, Eds. R.O. Ritche and E.A. Starke Jr. EMAS (U.K.), Vol. III, 1987, pp. 1439–1448

44. Lü P.M., Jiao S.J.: *An improved method of calculation the peak stress distribution for a broad-band random process*, Fatigue and Fracture of Engineering Materials and Structures, Vol. 23, 2000, pp. 581–586

45. Lutes L.D., Larsen C.E.: *Improved spectral method for variable amplitude fatigue prediction*, Journal of Structural Division, Vol. 116, No. 4, 1990, pp. 1149–1164

46. Łagoda T., Macha E.: *Transformation of stress frequency in criteria for multiaxial random fatigue*, Fracture Behaviour and Design of Materials and Structures, (ECF 8), Ed. D. Firrao, EMAS (U.K.), Vol. III, 1990, pp. 1311–1316

47. Łagoda T., Macha E., Achtelik H., Karolczuk A., Niesłony A., Pawliczek R.: *Multiaxial Random Fatigue of Machine Elements and Structures, Part IV, Energy Approach to Fatigue Life Including the Stress Gradients*, Studies and Monographs 139, Technical University of Opole ,Opole 2002, 102 ps, (in Polish)

48. Łagoda T., Macha E., Dragon A., Petit J.: *Influence of correlations between stresses on calculated fatigue life on machine elements*, International Journal of Fatigue, Vol. 18, No. 8, 1996, pp. 547–555

49. Łagoda T., Macha E., Niesłony A.: *Fatigue life calculations by means of the cycle counting and spectral methods under multiaxial random loading*, Fatigue and Fracture of Engineering Materials and Structures, Vol. 28, 2005, pp. 409-420

50. Łagoda T., Macha E., Niesłony A., Müller A.: *Fatigue life of cast irons GGG40, GGG60 and GTS45 under variable amplitude tension with torsion*, The Archive of Mechanical Engineering, Vol. XLVIII, 2001, pp. 55–69

51. Łagoda T., Macha E., Niesłony A., Müller A.: *Comparison of calculation and experimental fatigue lives of some chosen cast irons under combined tension and torsion*, Proc. ECF13, Application and Challenges, San Sebastian 2000 (Spain), Eds. M.Fuentes et al., Elsevier, Abstract Volume pp. 256–256, CD-Rom, 6 ps

52. Łagoda T., Macha E., Pawliczek R.: *The influence of the mean stress on fatigue life of 10HNAP steel under random loading*, International Journal of Fatigue, Vol. 23, No. 4, 2001, pp. 283–291

53. Macha E.: *Mathematical Models of the Life to Fracture for Materials Subjected to Random Complex Stress Systems*, Scientific Papers of the Institute of Materials Science and Applied Mechanics of Wroclaw Technical University, No. 41, Wrocław 1979, 99 ps, (in Polish)

54. Macha E.: *Generalization of Strain Criteria of Multiaxial Cyclic Fatigue to Random Loading*, VDI Reihe 18, Nr 52 VDI-Verlag, Dusseldorf 1988, 102 ps

55. Macha E.: *Simulation investigations of the position of fatigue fracture plane in materials with biaxial loads*, Mat. –wiss. u. Werkstofftech., Nr 20, 1989, Teil I, Heft 4/89, pp. 132–136, Teil II, Heft 5/89, pp. 153–163

56. Macha E.: *The spectral method of fatigue life calculation under random multiaxial loading*, Physicochemical Mechanics of Materials, Vol. 32, No. 3, 1996, pp. 86–96

57. Madsen H.O., Krenk S., Lind N.C.: *Methods of Structural Safety*, Prentice-Hall, 1986

58. MATLAB Function Reference: *Volume 2: F – O*, Copyright 1988–2002 by The MathWorks, Inc., Version 6, 826 ps

59. Miles J.: *On the structural fatigue under random loading*, Journal of Aeronautical Science, Vol. 21, No. 11, 1954, pp. 753–762

60. Morrill J.H., Achatz T., Khosrovaneh A.: *An application for fatigue damage analysis using power spectral density from road durability events*, SAE Technical Paper Series, No. 980689, Reprinted from: Advancements in Fatigue Research and Applications (SP- 1341), 1998, pp. 55–62

61. Morrow J.D.: *The effect of selected subcycle sequences in fatigue loading histories*, Random Fatigue Life Prediction, ASME Publications, PVP 72, 1986, pp. 43–60

62. MSC/FATIGUE User's Guide: *Vibration Fatigue Theory*, MSC.Software Corporation, 2002, 1360 ps

63. Naess A., Yim S.C.S.: *Stochastic response of offshore structures exited by drag forces*, Journal of Engineering Mechanics, Vol. 122, No. 5, 1996, pp. 442–448

64. Nagode M., Fajdiga M.: *A general multi- modal probability density function suitable for the rain- flow ranges of stationary random processes*, International Journal of Fatigue, Vol. 20, No. 3, 1998, pp. 211–223

65. NASA Preferred Reliability Practices: *Spectral Fatigue Reliability*, Johnson Space Center, Guideline No. GD- AP- 2303, 1993, pp. 1–5

66. Niesłony A., Macha E., Łagoda T.: *Generalization of spectral method of fatigue life determination under multiaxial random loading with the application of strain criteria*, Engineering Machines Problems, Vol. 20, 2002, pp. 119–126, (in Polish)

67. Oritz K., Chen N.K.: *Fatigue damage prediction for stationary wideband processes*, Proc. Fifth Int. Conf. on Application of Statistic and Probability in Soil and Struct. Engrg, 1987

68. Oziemski St.: *Analysis of Spectrum of Loading for Fittings of Excavators*, Monographs CPBP 02.05., Warsaw University of Technology, Warsaw 1990, ps 59

69. Perruchet C., Vimont P.: *Résistance à la fatigue des matériaux en contraintes aléatoires*, L'Aéronautique et l'Astronautique, No. 47, Vol. 4, 1974, pp. 71–81

70. Petrucci G., Zuccarello B.: *On the estimation of the fatigue cycle distribution from spectral density data*, Proc. Inst. Mech. Engrs, Vol. 213, Part C, 1999, pp. 819–831

71. Pierson W.J., Moskowitz L.: *A proposed spectral form for fully developed wind seas based on the similarity theory of SA Kitaigorodskii*, J. Goephys Res, Vol. 69, No. 24, 1964, pp. 5181–5190

72. Pitoiset X., Kernilis A., Preumont A., Piéfort V.: *Estimation du dommage en fatigue multiaxiale de structures métalliques soumisis à des vibrations aléatoires*, Revue Française de Mécanique, 1998, pp. 201–208

73. Pitoiset X., Preumont A.: *Spectral methods for multiaxial fatigue analysis of metallic structures*, International Journal of Fatigue, Vol. 3, 2000, pp. 541–550

74. Pitoiset X., Preumont A., Kernilis A.: *Tools for multiaxial fatigue analysis of structures submitted to random vibrations*, European Conference on Spacecraft Structures, Materials and Mechanical Testing, Braunschweig, Germany, 4–6 November 1998, (pdf version)

75. Pitoiset X., Rychlik I., Preumont A.: *Spectral methods to estimate local multiaxial fatigue failure for structures undergoing random vibrations*, Fatigue and Fracture of Engineering Materials and Structures, Vol. 24, 2001, pp. 715–727

76. Pitoiset X., Rychlik I., Preumont A.: *Spectral formulation of multiaxial high-cycle fatigue criteria for structures subjected to random load*, 6th Int. Conf. on Biaxial/Multiaxial Fatigue & Fracture, Ed. Manuel Moreira de Freitas, Vol. I, Lisboa, Tech. Univ. Portugal, 2001, pp. 297–304

77. Pook L.P.: *Spectral density functions and the development of Wave Action Standard History (WASH) load histories*, International Journal of Fatigue, Vol. 11, No. 4, 1989, pp. 221–232

78. Preumont A., Piéfort V.: *Prediction random high- cycle fatigue life with finite elements*, Journal of Vibration and Acoustics, Vol. 116, 1994, pp. 245–248

79. Rajcher V.L.: *Gipoteza spektralnogo sumirovania i jejo primenenia dla opredelenia ustalostnoj dolgovečnosti pri dejstvii slučajnych nagruzok*, Trudy CAGI, Č. 1134, Moskwa 1969, 3–38, (in Russian)

80. Rice S.O.: *Mathematical analysis of random noise*, Bell System Technical Journal, 23, 1944, pp. 282–332 and 24, 1945, pp. 46–156, or in: Wax N. Selected Papers on Noise and Stochastic Processes New York, 1954

81. Rychlik I.: *A new definition of the rainflow cycle counting method*, International Journal of Fatigue, Vol. 9, No. 2, 1987, pp. 119–121

82. Rychlik I.: *Rainflow cycles in gaussian loads*, Fatigue and Fracture of Engineering Materials and Structures, Vol. 15, No. 1, 1992, pp. 57–72

83. Rychlik I.: *On the 'narrow- band' approximation for expected fatigue damage*, Probabilistic Engineering Mechanics, Vol. 8, 1993, pp. 1–4

84. Sakai S., Okamura H.: *On the distribution of rainflow range for Gaussian random processes with bimodal PSD*, JSME International Journal, Series A, Vol. 38, No. 4, 1995, pp. 440–445

85. Sarkani S., Kihl D.P., Beach J.E.: *Fatigue of welded joints under narrowband non- Gaussian loadings*, Probabilistic Engineering Mechanics, Vol. 9, 1994, pp. 179–190

86. Sarkani S., Michaelov G., Kihl D.P., Beach J.E.: *Fatigue of welded steel joints under wideband loadings*, Probabilistic Engineering Mechanics, Vol. 11, 1996, pp. 221–227

87. Schütz D., Klätschke H., Steinhilber H., Heuler P., Schütz W.: *Standardized Load Sequences for Car Wheel Suspension Components*, LBF- Report No. FB-191, Darmstadt, 1990, 32 ps

88. Serensen S.V., Kogayev V.P.: *Fatigue life of machines components with regard to failure probability under nonstationary variable loading*, Vestnik Masinostroenia, 1, 1966, pp. 7–12, (in Russian)

89. Signal Processing Toolbox: *User's guide*, Copyright 1988–2002 by The Math-Works, Inc., Version 6, 1036 ps

90. Sobczyk K., Spencer B.F.: *Random Fatigue: From Data to Theory*, Academic Press, Inc., New York, 1992, ps 240

91. Sobczykiewicz W.: *Design of Work Attachment of Earth Moving and Construction Machines Against Fatigue Failure with Respect to Manufacturing Process and Operation Conditions*, Monographs CPBP 02.05., Warsaw University of Technology, Warsaw 1990, s. 99, (in Polish)

92. Sonsino C.M.: *Limitation in the use of RMS- values and equivalent stresses in variable amplitude loading*, International Journal of Fatigue, Vol. 11, No. 3, 1989, pp. 142–152

93. Sonsino C.M., Kaufmann H., Grubišić V.: *Transferability of material data for the example of random loaded forged truck stub axle*, SEA Technical Paper, No. 970708, 1997, pp. 1–22

94. Sun J.Q., Wang X.: *Fatigue analysis of non- linear structures with von Mises stress*, Journal of Sound and Vibration, Vol. 245, No. 5, 2001, pp. 947–952

95. Szala J.: *Hypotheses of Fatigue Damage Accumulation*, University of Technology and Agriculture, Bydgoszcz 1998, ps 175, (in Polish)

96. Symbolic Math Toolbox: *User's guide*, Copyright 1993–2000 by The Math-Works, Inc., Version 2, p. 266

97. Tang J., Ogarevic V., Tsai C.- S.: *An integrated CAE environment for simulation- based durability and reliability design*, Advances in Engineering Software, Vol. 32, 2001, pp. 1–14

98. Tovo R.: *A damage- based evaluation of probability density distribution for rain-flow ranges from random processes*, International Journal of Fatigue, Vol. 22, 2000, pp. 425–429

99. Tovo R.: *Cycle distribution and fatigue damage under broad-band random loading*, International Journal of Fatigue, Vol. 24, 2002, pp. 1137–1147

100. Tunna J.M.: *Fatigue life prediction for gaussian random loads at the design stage*, Fatigue and Fracture of Engineering Materials and Structures, Vol. 9, No. 3, 1986, pp. 169–184

101. Welch P.D.: *The use of fast fourier transform for the estimation of power spectra: a method based on time averaging over short, modified periodograms*, IEEE Trans. Audio Electroacoust. Vol. AU–15, 1967, pp. 70–73

102. Winterstein S.R.: *Nonlinear vibration models for extremes and fatigue*, ASCE, Journal of Engineering Mechanics, Vol. 114, No. 10, pp. 1772–1790

103. Wirsching P.H., Light M.C.: *Fatigue under wide band random stresses using rainflow method*, Journal of Structural Division, Vol. 106, No. ST7, 1980, pp. 1593–1607

104. Zienkiewicz O.C., Taylor R.L.: *The finite element method: solid and fluid mechanics, dynamics and non-linearity*, Vol. 2, 4-th ed., McGraw-Hill, London, UK,1991

105. Xu Y.L.: *Fatigue damage estimation of metal roof cladding subjected to wind loading*, Journal of Wind Engineering and Industrial Aerodynamics, Vol. 72, 1997, pp. 379–388

Index